I0039978

KNOWLEDGE
AND
DISCOVERY

Jeremy Y. Ng

KNOWLEDGE AND DISCOVERY

A Strategy Manual for Undergraduate Students Seeking Research Experience in the Natural and Clinical Sciences

Jeremy Y. Ng

Foreword by Dr. Smita Halder

First Edition

Knowledge and Discovery: A Strategy Manual for Undergraduate Students Seeking Research Experience in the Natural and Clinical Sciences

First Edition

Copyright © 2013 by Jeremy Y. Ng

All rights reserved. No part of this publication may be reproduced, distributed, or transmitted in any form or by any means, including photocopying, recording, or other electronic or mechanical methods, now known or hereafter invented, without the prior written permission of the publisher.

Published in Canada, September 2013

ISBN 13: 978-0-9920840-0-4 (paperback)
ISBN 13: 978-0-9920840-1-1 (Ebook, mobi)
ISBN 13: 978-0-9920840-2-8 (Ebook, epub)

ISBN 10: 0-9920840-0-8 (paperback)
ISBN 10: 0-9920840-1-6 (Ebook, mobi)
ISBN 10: 0-9920840-2-4 (Ebook, epub)

Library and Archives Canada Cataloguing in Publication

Ng, Jeremy Y., 1991-, author
Knowledge and discovery : a strategy manual for undergraduate students seeking research experience in the natural and clinical sciences / Jeremy Y. Ng. -- First edition.

Includes bibliographical references and index.
Issued in print and electronic formats.
ISBN 978-0-9920840-0-4 (pbk.).--ISBN 978-0-9920840-1-1 (mobi).--ISBN 978-0-9920840-2-8 (epub)

1. Undergraduates--Research--Handbooks, manuals, etc. 2. Education, Higher--Research--Handbooks, manuals, etc. 3. Science--Research--Handbooks, manuals, etc. 4. Medicine--Research--Handbooks, manuals, etc. I. Title.

LB2326.3.N45 2013   378.0072   C2013-905838-9
                                         C2013-905839-7

For Gordon and Betty Rainbow

Ordering

Additional paperback copies of this strategy manual can be purchased from Amazon.com or the local Amazon.com, Inc. retail website of your country.

Additional EBook copies of this strategy manual can be purchased in .mobi format from Amazon.com or the local Amazon.com, Inc. retain website of your country, and in .epub format from the Apple iBookstore and Barnes & Noble NOOK.

Contact and Support

For further reading, please visit the author's website at:
http://researchyourself.wordpress.com

Please support the distribution of this book through Facebook®! Please visit:
Facebook.com/KnowledgeandDiscovery

Any inquiries regarding this book should be directed to the author Jeremy Y. Ng through email at:
ng.jeremyy@gmail.com

Legal Disclaimer

This book is sold with the understanding that both the author and publisher are not offering it as legal, accounting, or other professional services advice. While the best efforts have been used in preparing this information provided in this publication, both the author and publisher assume no responsibility for errors, omissions, or contrary interpretation of the subject matter herein, and make no representations or warranties of any kind and assume no liabilities of any kind with respect to the accuracy or completeness of the contents and specifically disclaim any implied warranties of merchantability or fitness of use for a particular purpose. Neither the author nor the publisher shall be held liable or responsible to any person or entity with respect to any loss or any physical, psychological, emotional, financial, or commercial damages, including, but not limited to, special, incidental, consequential or other damages caused, or alleged to have been caused, directly or indirectly, by the information contained herein.

CONTENTS

Knowledge and Discovery

FOREWORD
By Dr. Smita Halder

It was a great honour to be asked by Jeremy to write the foreword to this unique and vital book that encourages students from all natural or clinical science backgrounds to involve themselves in undergraduate-level research, in order to enrich their university experience and gain practical and applicable real-world skills.

I have known Jeremy since the third year of his undergraduate career when he first approached me expressing his interest in gaining clinical experience through shadowing me in gastroenterology. In his fourth year, he undertook his senior thesis under my supervision and focussed on assessing the prevalence of anxiety and depression in inflammatory bowel disease patients attending the university clinic, and upon his graduation we have maintained contact. Jeremy has always worked with diligence, maturity and a passion to produce excellence, so when he told me about the book I knew that he would only complete this endeavour under the condition that he put forth his best efforts and sincerest dedication. As a result, it is clear to me that he has written something for undergraduates that I believe is truly relevant and valuable.

The idea of offering research experience to undergraduates has been around now for many years, however, recently due to increased advocacy and opportunity provided by universities and principal investigators, it has become a recommendation, if not even a requisite, for the many students wishing to pursue further higher education. As we live in a world shaped by competition, it is no surprise that countless students make the attempt to gain research experience every year, however, only those who have knowledge of the best application strategies succeed in landing a position.

It is my hope that this book serves to grow interests in gaining research experience by sharing the direct benefits that can be gained in today's undergraduate setting, as follows:

1. Gaining research experience can (and often will) make the difference between whether a professional or graduate school application is successful.
2. A summer research job can pay off an entire year of undergraduate tuition.
3. Research experience can help students to gain a referee who is a high-profile, well-respected and well-accomplished academic scholar.
4. Research experience will teach a student real-life, practical application skills that cannot be gained by taking an undergraduate course.
5. Exposure to research will change the way students think, learn, understand, and perhaps even the way they lead their lives, all for the better.

In *"Knowledge and Discovery: A Strategy Manual for Undergraduate Students Seeking Research Experience in the Natural and Clinical Sciences"*, Jeremy has taken great care to deconstruct every aspect of undergraduate research from beginning to end. By employing the strategies contained within this manual, not only does the student set themselves apart from their peers who simply focus their efforts on reading a textbook needed to pass their next exam, but they are also opened up to a world of benefit and opportunity that has the potential to positively impact their future goals, education, professional career, and even shape the way one understands the world around them.

To my knowledge, this is the very first book that provides comprehensive coverage of the undergraduate research application process, with the added advantage that it is written from the perspective of a student. It is for this reason that I believe this book will find a far and wide-reaching undergraduate student audience, and I shall be recommending it to any undergraduate student seeking research in the natural or clinical sciences as this book is truly a tool of encouragement. I know that Jeremy has made the best efforts to produce a strategy manual that is not only concise, but also all inclusive, in the hope that the advice contained herein will provide undergraduates everywhere with academic success and inspiration.

Dr. Smita Halder, MBChB, MRCP, MRes, PhD
Assistant Professor, Division of Gastroenterology, Department of Medicine, McMaster University
Hamilton, Ontario, Canada

Biography of Dr. Smita Halder

Dr. Smita Halder studied medicine at Cambridge and Edinburgh Universities. She returned to her home city of Manchester to train in gastroenterology, and in 2002 was appointed an MRC fellow in Health Sciences Research. After completing her PhD, Smita pursued a postdoctoral fellowship at Mount Sinai Hospital in Toronto. Initially intending to stay for one year, she has now made Canada her home and is an Assistant Professor in the Department of Medicine at McMaster University in Hamilton, Ontario. She has a busy clinical practice and subspecialises in the management of inflammatory bowel disease. Her research interests include the epidemiology of IBD and the delivery of effective health care in chronic disease. Outside of work, Smita's main passion is travelling the world, which has temporarily been curtailed since the arrival of her baby boy.

PREFACE
By Jeremy Y. Ng

Imagine partaking in an activity that allows you to gain experience beyond the traditional framework of an undergraduate career, where instead of only acquiring knowledge, you are creating it, and instead of only learning about discoveries, you are making them. This is what it is like to be an undergraduate researcher.

This opportunity is very unique as many undergraduates who involve themselves in research develop valuable practical skills that cannot be learned through any university course, priming them for future success. It is for this sole reason, that I encourage you to experience research at the undergraduate level, for it is always the case that there will be little to lose and far more to gain.

My inspiration to write this strategy manual arose from the fact that, upon much searching, I found that there are virtually no books that comprehensively explain how undergraduates can gain research experience. As a result, I know this material will benefit many students because I have written a book that I wish I could have read before gaining my first research experience.

I would argue that research is best described by the creation of knowledge and the requisite to discovery, hence the title of this book. After ambitiously challenging myself to write two pages of this strategy manual each day, following the conclusion of my successful undergraduate research career, I am pleased to present you with *"Knowledge and Discovery: A Strategy Manual for*

Knowledge and Discovery

Undergraduate Students Seeking Research Experience in the Natural and Clinical Sciences". In this manual, you will find 12 comprehensive chapters outlining everything you need to know about undergraduate research, an activity that arguably has the potential to provide you with more academic success and advancement opportunity than any other undergraduate pursuit.

My ability to write this strategy manual is attributed to the numerous research experiences that I have personally gained over the past four years of my undergraduate career spanning both the natural and clinical sciences. Not only have I worked alongside seven different professors including distinguished research chairs and clinician-scientists, but I have also had the opportunity to use my experiences in research to pay for my entire undergraduate degree, acquire research grants of my own, gain academic jobs in teaching, develop relationships with invaluable mentors, and gain entrance into the graduate school program of my choice. As a result, you can be assured that the provision of the strategies contained within this book stem from the fact that I have a successful track record when it comes to gaining research experience at the undergraduate level.

This book has been specifically written so that it can be read sequentially or separately based on your level of experience and reading preference. The beginning chapters provide you with a thorough understanding of the structure of research and professorships, the benefits students may gain from this activity and the types of research positions available to undergraduates. The vast majority of this strategy manual provides an in-depth and highly comprehensive look into the application process, inclusive of all the information you need to find an appropriate research group, construct and submit your application, and prepare for the interview with potential supervisors. Lastly, the final chapters explain how you can successfully optimize your research experiences, maintain your professional relationships, and continue to advance in research throughout the rest of your undergraduate career. Through reading this book and using the advice within it, I have full confidence that you will gain a competitive advantage over your peers, while bypassing the vast majority of common mistakes that many undergraduates typically make along the way.

I have purposefully written this strategy manual to accommodate a wide-reaching audience. Whether you are the high school student who knows little about the academic research environment, the seasoned final-year undergraduate seeking opportunities for advancement, or a student who fits into a category somewhere in between, there is bound to be something valuable for you to gain. Additionally, if you are a faculty member, graduate student or teaching assistant, I highly recommend sharing this strategy manual with any undergraduate students who express their interest in research.

As a student, research experience will teach you how to think creatively and critically, promote your academic development, allow you to learn about your future career options, and perhaps even provide you with the benefit of reaping rewards that you never thought were possible. It is my sincere hope that this book piques the interest of students from all natural and clinical science backgrounds who otherwise would not pursue undergraduate research, but also encourages and inspires those who already have an established interest or who carry prior experience. Finally, remember that there is nothing more powerful than the creation of knowledge and the provision of an opportunity that holds the potential to make a positive impact in the world around you.

I wish you only the best in your academic research endeavours.

Jeremy Y. Ng, BSc
Primary Author and Editor
Knowledge and Discovery: A Strategy Manual for Undergraduate Students Seeking Research Experience in the Natural and Clinical Sciences
Toronto, Ontario, Canada

Acknowledgements

I wish to express my gratitude to both family and friends who witnessed the process involved in writing this book, providing encouragement, support and feedback and assisted in the editing and proofreading process.

My special thanks goes to the professors who served as my mentors and/or supervisors, both in and out of research, during my time as an undergraduate student, who were a help, an encouragement, and an inspiration to me, each at different times and in different ways.

Dr. D. Boreham
Ms. U. Chauhan
Dr. M. Dittrich
Dr. J. Draper
Dr. S. Halder
Dr. T. Kislinger
Dr. A. Neville
Dr. E. Ratcliffe

In particular, I also wish to acknowledge my undergraduate senior thesis supervisor, Dr. S. Halder, an inspirational physician and teacher, for kindly writing the Foreword to my book.

Finally, I must recognize Mr. G. Rainbow, a life-long friend and invaluable mentor, for providing his contributions to this book, including the provision of helpful suggestions, new ideas, and assistance in proofreading.

This is not at all to say that the aforementioned are in favour or support of all contained in the following pages for that responsibility is mine, and mine alone.

Last and not least, I apologize sincerely and ask for forgiveness from all those who have been with me over the course of this journey and whose names I have failed to mention.

CHAPTER 1
Undergraduate Research in Context: An Introduction

An Introduction to Undergraduate Research

If you were like me, you had a very minimal understanding of what the "university experience" entailed before you began your undergraduate degree. Being the first-born child or being the first person in your family to attend university only makes this task more difficult. Beginning an undergraduate career is a significant milestone in any student's life. For the first time, a student will be immersed into a whole new educational setting and lifestyle, often after leaving home to live without their family, friends and social network. While this can have both negative and positive imprints, for anyone who has been to university, they would agree that their undergraduate education was life-changing and unlike any other opportunity they have ever experienced in their life.

Within the first couple weeks of your freshman year, you will have had the opportunity to attend all of your classes and perhaps even meet with some of your professors, yet among many high school and undergraduate students, the role of a professor, beyond the person they know as giving the lecture in the auditorium, is largely unknown. Each professor is (or was) involved in research, and has contributed to the field of academia in some way. In fact, to become a professor, you need to become a knowledgeable expert in a certain field. To do this, professors recruit students to help them with their work, including undergraduates (and occasionally, very keen high school students).

Purpose and Aims

In this book, I will attempt to share some of my experiences with you, while providing examples of how research experience in both the natural and clinical sciences at the undergraduate level opened many doors of opportunity in my education and my life. It is my hope that I will be able to convey the importance of gaining research experience during your undergraduate career through this book. If you are a high school student planning to attend university or an undergraduate student either uncertain of gaining or actively searching for research experience, it is my hope that you will find the contents of this book both motivating and helpful.

"So, what exactly is research, and how can I get involved?" This is a common question asked by most students upon first hearing about undergraduate research opportunities. Without being too technical, research can be defined as "an investigation with the purpose of acquiring new knowledge". This investigation can manifest itself in many forms, and can span many different fields. From natural and clinical sciences, to humanities and social sciences, to engineering sciences and everything else in between, all the knowledge that has accumulated into a pile of pages known as your course textbook has been acquired through research, and only research alone. Research fundamentally provides us with the answers to our most sought after (and often most difficult) questions, and it is for this reason that it is so vitally important to advancing any field.

The primary aim of this book is to help you to gain an interest in, and prepare for an undergraduate research experience, which I will do by sharing some of my own personal experiences in research, as well as the advice that I was provided with by my professors and their graduate and undergraduate students.

Most undergraduates, especially those just beginning their university experience, react hesitantly when asked whether they wish to gain research experience. This is quite normal, and I too, must admit I felt the same way when I was in my first year. Towards the end of the first year, a friend of mine asked me what I was planning to do during the upcoming summer, knowing that I had no solid plans. She suggested that I try to apply for a research position with a university professor. I was very hesitant, as after all, I knew nothing about research. All I knew was that professors already had second, third and fourth year students applying to their research group, and that my application likely stood no chance over what I believed to be the more competent competition. Despite my doubts, in the end I did spend my Reading Week of that year applying for research positions, eventually landing my first job.

Unfortunately, this story is far from the case with many other students. Too many students never hear about the opportunities in research, and even if they do they hold onto the sense of doubt that I experienced, and often lack the confidence to inquire about how to get involved in the research experiences that are being offered by their own university. Experience has taught me that students can always make an unlimited amount of excuses for their sense of doubt, and the most common one regarding research experience can be summarized by the following question: "What if research is not interesting?"

Unfortunately, the sad reality is that many universities struggle to emphasize the reasons why research is exciting and beneficial to undergraduates, however, no individual or group should be blamed in particular, as this is an inherently difficult task. Many students would rather join what is familiar to them, such as a sports team or a student organization that suits their already-developed interests and hobbies. Often, research opportunities are advertised in places with low student traffic, such as on a corkboard or on an obscure and difficult-to-access department website. It is much less common that a research opportunity is announced by the teaching professor in a lecture hall, and even if this does happen, while a few keen students will apply immediately, the vast majority of students will ignore the announcement. Despite all of this, I still stand by the fact that research experience is invaluable because the skills learned are highly applicable to a student's further education and future career.

Personal Experiences in Research

Before I provide you with the strategies you need to gain research experience, I first wish to share my own experiences with you. From these experiences, I will then be able to draw out relevant examples throughout the rest of this strategy manual.

Following high school, I enrolled in the life sciences program at McMaster University in Hamilton, Ontario, Canada, graduating with a Bachelor of Science in biology. One of the things that attracted me to McMaster University was the fact that the ongoing research was broad, so that I could explore different fields and find what interested me most.

I gained my first research experience (and first job, for that matter) in the summer directly following my first year, at a biogeochemistry laboratory at the University of Toronto Scarborough. This job was relatively elementary in nature, in fact, because my professor had been hired just that same year, the majority of my tasks included organizing, labelling and washing glassware. This was not an ideal research experience, but it got my foot in the door, and it taught me much about what to look out for when searching and accepting research positions. To this day, I am thankful for this research experience, since many professors I had applied to were not even willing to consider my application because of my lack of experience, for had I applied knowing what I know today, the opportunities I would have been offered would have likely been more plentiful.

My second research experience took me to the University Health Network, where I worked alongside a professor of medical biophysics. Here I was accepted as one of 50 students to a competitive undergraduate summer studentship. Despite all the other research experiences to follow, this position probably provided me with the most experience in how the research world worked, and for the first time, I was able to sharpen my skills in basic laboratory techniques, read scientific literature, and practise my academic writing.

My next research experience was gained at the Stem Cell and Cancer Research Institute at McMaster, as one of the institute's few undergraduate volunteers. Throughout my third year, I completed many cell imaging experiments, and then later completed a research course with my supervisor, in which I learned various additional molecular laboratory techniques.

During the summer following my third year I was employed by the McMaster Institute for Applied Radiation Sciences, where I was awarded two undergraduate research grants. It was during this time that I trained in animal handling, to perform experiments involving rodents.

Finally, in my fourth year I completed both my senior thesis and research placement with a professor of medicine at McMaster University whom I had shadowed during the summer. This was my first research experience in the clinical sciences, which was unique, as I was able to interact with and interview patients, collect survey data and learn about epidemiological research.

I am currently a graduate student at the Leslie Dan Faculty of Pharmacy at the University of Toronto, where my research focusses on complementary and alternative medicine, a field which has fascinated me since grade school.

You may be wondering how my graduate program relates to the experiences I gained during my undergraduate career, as after all, it appears that none of my research had anything at all to do with complementary and alternative medicine. This could not be further from the truth, however, as a field like complementary and alternative medicine is broad in nature, like many other fields in health.

My undergraduate experiences have helped me to better understand studies that investigate complementary and alternative therapies in a laboratory setting, whether they are being tested on human cell lines or animal models. Furthermore, it is very common for complementary and alternative medicine therapies to be tested in a clinical setting requiring the compliance of patients. It is here that my experience in patient interaction and interviewing as well as data collection and epidemiology became useful. Finally, a standard requirement in graduate school is to possess strong academic writing skills, which I gained from writing abstracts, posters, grant applications and laboratory reports throughout each of my prior research experiences.

The most interesting thing, however, is that I never once planned to learn any of these exact research experiences purposefully. I had carefully applied to different research groups that interested me, and sought to join a research team that could provide me with a skill I was looking to learn, but then the rest fell into place. It is for this sole reason, that I encourage you to experience research at the undergraduate level, for it is almost always the case that there will be little to lose and far more to gain.

Diversity in Research

The range of research opportunities that an undergraduate can choose from is extremely broad and limitless. One reason why students harbour a poor perception of research is because they make the assumption that academic research is similar to their high school or first year "laboratory experiments".

Like any other student, I also participated in these mandatory "laboratory experiments". As I recall, this required students to work at a crowded lab bench, alongside perhaps a group of unknowledgeable or uninterested classmates. In addition, the goal was to complete a boring, tried, tested and true experiment in a rushed environment, all the while using broken or abused equipment older than the students themselves. As a result, I can understand how this fuels a student's negative perception of academic research even before they begin their undergraduate degree, however, believe me when I say that this pre-conceived idea could not be further from the truth.

Instead, imagine that you are able to explore any unanswered research question you have ever wanted to investigate. If you are a student in the natural sciences, perhaps you wish to learn more about the ecology of a certain species, or perhaps understand how a certain plant grows and adapts to its climate. If you are a student in the social sciences, perhaps you wish to understand why so many students struggle with depression, or why certain age groups choose to consume more alcohol. If you are a clinical sciences student, perhaps you want to know why certain human populations are more prone to developing brain cancer. The possibilities are virtually endless, and at any large (or even medium) sized university, there is often someone who does research in or relating to your area of interest.

Some students may not know what their interest is, and instead, choose to gain an interest by exploring different research groups and the projects being offered. Either way, I encourage you to picture this as what the academic research environment can offer you. Beyond the fact that you are bound to find something interesting, you will often have access to state-of-the-art research equipment, and a very knowledgeable mentor and students who share a genuine interest in conducting experiments. If I can convince you to adopt this new mindset before considering whether you want to gain research experience, I would hope that this would provide you with the encouragement you need to apply for and become engaged in a research experience that interests you.

Addressing the Infamous Catch-22

If this still does not convince you, consider that having research experience often provides students with a huge advantage when pursuing their post-undergraduate goals. If you are like most students, you likely have aspirations to gain entrance to a professional or graduate school, or begin a career working for the government or a large company. As I will explain in the next chapter, it pays to have research experience to provide you with that competitive edge.

Fundamentally, obtaining a summer job or volunteer position in the research world at the undergraduate level is similar to any other job suitable for a student with respect to experience. To gain research experience, you need to get research experience. I was fortunate enough to gain a summer research experience directly following my first year of university, without having any prior research experience. A select few of my friends and classmates also succeeded, but for the majority, research experience was not something that they would gain until their third or fourth year.

Undoubtedly, this is one of the most frustrating and challenging catch-22 situations students at the undergraduate level face, yet from a professor's standpoint it makes complete sense. It is not advantageous for a supervisor to invest their time, energy and resources in a student who is inexperienced, unknowledgeable and untested in their skills. Any professor, given the amount of effort it takes to obtain research funding, would rather hire someone who has both solid theoretical knowledge and practical skills required to perform the necessary research tasks.

For the inexperienced student, this means putting in the time and effort to show that you are a stellar candidate. It is interesting to see how undergraduates define a "stellar candidate", however, as this often differs from a professor's definition. Students often believe that good grades are the sole currency required in gaining entry to an undergraduate research assistantship. As a result, those with low grades neglect to apply because of this belief, but as I will explain later, I cannot express any more clearly how far this is from the truth. While grades play a small part in the application process, there are many other factors within your greater control that play an even more important role. I encourage you to not give up, even if you do not believe that you are a stellar student academically, as gaining research experience is very dissimilar to gaining acceptance to a highly competitive academic program.

There are virtually an unlimited number of positions available, and if you cannot secure a position with one professor, you can always move on to another. There are also no deadlines to apply, unless you are applying to a specialized research program, and you can almost always gain a research experience whenever you want one during your undergraduate career. Finally, always remember that it is never too late to gain research experience. Even if you are in your fourth year and graduating next month, with the proper knowledge, you can still gain a valuable experience in the summer following your graduation, thus preparing yourself for your future academic endeavours.

If you are still not convinced that gaining a research experience is beneficial during your undergraduate career, it is my hope that I will be able to convince you of its importance from a professional development perspective in the second chapter of this book where I explain the importance of undergraduate research. Even if you are convinced, I encourage you to read on as I discuss the numerous benefits associated with undergraduate research, some of which may be largely novel to undergraduates.

CHAPTER 2
The Importance of Undergraduate Research

Gaining the Competitive Edge

It is quite understandable why undergraduate research experience would be useful if students want to become professional researchers, however, for most students this is not their future career goal. Many students I encountered during my undergraduate degree often made the assumption that they did not require research experience only because they did not wish to pursue a profession in research. Unfortunately, in today's competitive undergraduate environment, that argument has very little value. The stark reality is that undergraduate research experience is sought after by professional schools, graduate schools, and even employers in the government or industry. The fact is that students with even one research experience have a significant advantage over students with none. Students holding research experience not only show interviewing committees that they have an interest in seeking the answers to difficult questions, but they also demonstrate that they are able to work in an environment that is both demanding and challenging.

Benefits for the Professional School Application

Many professional schools argue that gaining undergraduate research experience allows students to enrich their understanding of their university coursework and investigate real world applications of what they have learned from a textbook. Another reason why undergraduate research experience is valued by professional schools is because it allows students to hone their critical-thinking skills. Conducting any research experiment, especially in the laboratory environment requires concentration, careful planning, and most important of all, a specific intention for every step of the protocol.

Finally, whether you wish to become a physician, pharmacist, optometrist or dentist among many other professional school careers, your life will most likely be filled with life-long learning. My understanding of this comes from personal job shadowing experiences, as well as those of fellow classmates interested in attending professional school. Health care providers will always need to read up on the latest medical research, assess their findings and evaluate the validity of these studies that aim to cure diseases and improve treatments. All of this has direct application to how any health care provider chooses to evaluate and improve their quality of patient care. Ideally, if you wish to apply to a professional school, you should aim to gain some research experience, or better yet, join a research group relating to your own field. For example, if you wish to be an oncologist, you may want to consider joining a cancer research laboratory, as this shows that you have an early interest in the goals you wish to pursue.

Benefits for the Graduate School Application

If your goal is to attend graduate school as it was for me, then prior research experience may actually be a prerequisite. Graduate school can be summarized in three words: training in research. Students apply to graduate programs for many different reasons, as some may endeavour to become a university professor or lecturer, others may want a higher paying career in industry or government, and some may even use it as a stepping stone to gain a competitive edge when applying to a professional school. Finally, many students apply to graduate school because they really enjoyed the research aspect of their undergraduate career and want to begin to focus on a project and field which is most interesting to them. Regardless of what the reason may be for applying to graduate school, there is no doubt in my mind that you will need to gain some form of undergraduate research experience.

It is important to make the distinction between undergraduate and graduate studies as they are vastly different in nature. While an undergraduate degree can be completed by passing a certain number of courses, comprised of evaluations such as tests, assignments, essays and exams, there is inherently very little to no research requirement to earn a Bachelor of Science degree. This presents a problem to undergraduates who possess little research experience and who intend to further their education through a Master's or PhD. Either graduate degree, with few exceptions, require the student to not only become heavily involved in research, but often also manage or even lead multiple research projects, and eventually present their work at a scientific conference and publish their findings in an academic journal. For these reasons alone, many potential Master's supervisors distinguish between a strong applicant and a less strong applicant by evaluating their academic research accomplishments on their curriculum vitae. As a result, the more research experiences, research grants, and research presentations (or better yet, publications) you have acquired on your curriculum vitae as an undergraduate, the better chance you stand at being a supervisor's top applicant.

Research Experience Provides Financial Gain

The strength of your curriculum vitae, however, is not just important for those applying to graduate school or professional school. Beyond post-graduate education, possessing undergraduate research experience provides students with great benefits. For example, even if you choose not to further your education following the completion of your undergraduate degree, any potential employer would evaluate your resumé to identify your academic involvements before even considering inviting you for an interview. Thus, having research experience is a certain way to impress a potential employer in any job market hiring an undergraduate science major.

If you are just about to begin your undergraduate career, or are enrolled in your first or second year, gaining research experience can also have a very direct and positive impact on your achievements and successes even before you graduate.

A great example of this is a co-operative degree program. Undergraduate degrees that offer co-operative programs provide students with the advantage of integrating an entire term of real-world work experience in between the usual semesters of university coursework. This opportunity is sometimes even provided more than once throughout the duration of the degree. It is often the case that many co-operative students meet their future employers here, or at the very least build a network of professional relationships with great people. Eventually, the hope is that the student may return back to the company they completed a co-operative term with after graduation for full-time employment.

Unfortunately, the reality is that many of these programs are competitive, especially within the natural and clinical sciences. Once again, students face a catch-22, where they are required to demonstrate that they have experience in order to gain experience. Often, many co-operative degree options begin in the second or third year of the student's program. This means that students who apply are evaluated based on their performance during the first one or two years of their undergraduate careers. While grades are often the most important (or sole) factor in the decision making process, it is definitely advantageous to also carry prior research experience. This experience can be shared with the selection committee through your curriculum vitae or during the co-operative program interview.

If you can show the committee that you have been successful enough to have gained research experience in the past, you will likely also be perceived as a candidate who will have greater success in securing a co-operative job on your own because of your past research experience. This is also very important, as once students are accepted into a co-operative program, it is quite often the case that the most relevant co-operative jobs to the student's field will also be the most competitive jobs to obtain. As a result, students possessing prior research experience will stand out among all the applicants. The unfortunate reality is that some co-operative programs will only provide their students with minimal help in securing jobs because of a lack of student positions or due to the competitive nature of the field's industry, thus once again if you plan on being a co-operative student it is in your best interest to demonstrate that you have a competitive edge.

Some students decide that a co-operative undergraduate degree program is not what they want for various reasons, while many more programs do not offer any co-operative options. When I was deciding whether I wanted to pursue a co-operative program option following my first year of university, I decided against it. I am by no means saying this is the right choice for everyone, but I knew that I would have to spend an extra year to complete my undergraduate degree if I enrolled in the co-operative program. For this reason, it is important that you are aware of the length of the co-operative program that interests you, as some allow students to graduate in four years, while others add an additional year to your degree. Instead of enrolling in a co-operative program, I took full advantage of my summers between my undergraduate years, gaining paid research experience in this way instead, and I have no regrets about making this decision.

Whether you are enrolled in a co-operative program or not, there are additional benefits associated with undergraduate research experience worth mentioning. The first and most obvious is the financial gain associated with paid research experience. Unless you are incredibly wealthy, you will agree that the cost of tuition at university is both extremely expensive and debt inducing. Paid summer research jobs can offset this. For example, in one summer I earned enough money to pay off more than an entire year of my undergraduate tuition fees! In my opinion, gaining a summer job in research is great because you can simultaneously accomplish two goals such as gaining valuable experience that helped me to gain entry into graduate school while also reducing my burden of undergraduate tuition fee debt.

Academic Distinction in Research

In addition to financial gain, my research experiences have also provided me with research award opportunities, through undergraduate research grants. Undergraduates involved in any form of research, especially over a summer period, are often eligible to apply for some form of undergraduate research grants, which are awards that not only provide distinction to one's curriculum vitae, but also often provide or increase the student's salary. I will discuss the acquiring of undergraduate research grants in further detail in Chapter 11.

Yet another opportunity that arises from undergraduate research experience is the opportunity to contribute to the scientific community. Many research professors often encourage their students to write an abstract or research poster for submission to a scientific conference or academic symposium. Better yet, some undergraduates who manage to make significant findings within the duration of their research term become a co-author of a research publication. Again, I will discuss this topic in greater detail in Chapter 11 of this book.

Creating an Academic Network

Many of the aforementioned benefits of undergraduate research I have mentioned are very much based on increasing your academic accomplishments. In essence, this means that upon gaining research experience, you are now able to add a few entries to your curriculum vitae. While this is not a bad thing (in fact, it is necessary in today's competitive world), I would never encourage a student to make this their primary goal in their undergraduate career. This is a destructive mindset because it encourages you to make commitments to tasks that are of little interest or value to you. Instead, you should focus on networking, the biggest opportunity research placements provide undergraduates.

Networking is often the only way to actually gain a professional relationship with a well-qualified, academic scholar in a field of your interest. A research experience not only provides you with a (typically) long-term relationship with a professor, it also gives you the opportunity to interact with graduate students and other faculty. Try your best to attend scientific conferences, symposiums or colloquia relating to your research position. This will allow you to broaden your knowledge of what is happening at the cutting edge of research in your particular field. As an undergraduate, you may need to make an effort to seek out when these events are being held, as invitations are often only privately circulated within department research staff and student circles. Sometimes a simple request to the department's secretary will allow for your email address to be added to the departmental mailing list that will periodically inform you of upcoming academic events and activities.

You may not necessarily understand the material being presented at these academic colloquia or scientific conferences, as I certainly did not understand all of the research jargon when I was an undergraduate. Yet, do not let this stop you from attending, as many of the faculty will acknowledge your interest to learn more about their field and be more than willing to answer any of your questions. If possible, ask questions about presenters' research field and career, as perhaps you may meet them again as a colleague or collaborator if you choose to progress further into this area of research.

One of the most valuable advantages of meeting those who are currently pursuing a higher level of education than you is that you can learn from them. Often, if you are working in a research group as an undergraduate you will be assisting with a small component of a graduate student's project. Graduate students can often relate to undergraduates more easily because they just finished their Bachelor's degree a few years ago. I often took this opportunity to ask graduate students what they were gaining from graduate school, and what interested them in the particular department and professor for whom they were working. It is here where you can learn about their successes (and failures) and decide for yourself whether this is an educational route that you wish to take in the future. It is important to ask graduate students what they would have done differently in both their undergraduate and graduate studies. Perhaps some graduate students wished that they had enrolled in a different program, worked for a different professor or acquired more research opportunities in their undergraduate career to gain a better understanding of different fields of research.

Take the opportunity, when possible, to also ask your professor about his or her career path. Some professors also hold a professional degree (i.e. a medical degree) in addition to holding a PhD. If you are interested in a professional school that your supervisor attended, ask them about what it was like to complete their professional degree. Often you can gain much more information from research supervisors, as opposed to simply shadowing a medical doctor, for example, because of the nature of research. While job shadowing may last only a week, you may interact with a professor multiple times during a research position, allowing you to grow your relationship with your supervisor.

Forming Professional Relationships

Undoubtedly, one of the biggest and best advantages to forming a long-term positive relationship with a research professor through a summer job or research placement is the opportunity to ask them to serve as your reference. Whether you apply to professional or graduate school, or for awards and scholarships, or even to your next undergraduate research job, you will at one point require a reference and/or a reference letter. For competitive programs, such as the graduate schools I applied to, a minimum of three professor references were required! This is a major reason why I encourage you to start your research career as early as possible during your undergraduate degree.

As an undergraduate at university (especially a large university), there is often very little opportunity to interact with any professor outside of research. Perhaps you could talk to your teaching professor after lecture, or ask good questions during their office hours (if they hold them), but either way, the time you will be able to spend will be minimal at best, and I highly doubt that you will have much success in gaining a long-term, meaningful professional relationship with this professor. Of course, there are other options, such as job shadowing a clinician-scientist, but these professors must actively attend to their practice so your interaction time in clinic is often limited. Furthermore, many clinician-scientists do not wish to take on observers as they see this as a disturbance to their patients. Finally, many fields outside the health sciences may not offer any job shadowing opportunities at all. Another strategy for gaining a professional relationship with a professor is to become their teaching assistant, as this is another good way to showcase your skills and competencies to a professor. This, however, is not easy during your first, second (or even third) year of university as contracts often state that teaching assistants must be final year undergraduates or graduate students. Thus, my main point here is that gaining a long-term professional relationship with a professor is best achieved in research early on in your undergraduate career.

Through any research experience, your professor will have the ability to serve as an excellent reference to you, because the nature of research provides them with the opportunity to evaluate your skills in critical thinking, leadership, teamwork, creativity, enthusiasm, dedication, and self-motivation, beyond just what they know about your courses and grades. Furthermore, if you complete a research placement course with your professor, he or she will serve as a reference who can dually comment on your aptitude for research as well as your academic performance.

Realize that a reference letter is often the only document you will never see before you submit your application. In other words, the outcome of this portion of your submitted application is in the hands of someone else. In addition, the reference letter is often one of the portions of the application that is closely scrutinized, and may make the difference between an acceptance and rejection once the deciding committee begins assessing the most competitive applicants. That being said, gaining a good referee is nothing to worry about if you have the correct approach. The key is to begin this process early.

Begin contacting and connecting with research professors before many of your peers, as you need to keep in mind that if you are aiming to gain acceptance to a professional or graduate program, you may require up to three (academic) references. Understand that forming relationships with professors takes time, and maintaining these relationships requires time as well, thus if you wish to be well-regarded by three professors, it is in your best interest to start your undergraduate research career as early as possible. I was fortunate enough to begin my research career early, so that when I was applying to graduate school, I was able to choose between six or seven professors at any given time to write my reference letters. This was simply because I had been very much involved in research since my first year and had maintained positive relationships with all the professors for whom I had worked. While most students will not form that many professional relationships with professors in their time as an undergraduate, it is quite realistic to assume that you could have three professors who know you well by the time you require a reference letter, if you put in the time, care and effort to start early.

Maintaining a relationship with a professor you no longer research with, can often be challenging for some students. I will discuss this further in Chapter 10, however, for now I think that I have provided you with enough information to convince you that research experience is extremely important for the undergraduate student in the natural or clinical sciences. In the next chapter, I will provide some important information about the structure of research at an academic institution, and also some advice on how you can integrate your research experiences into an already hectic undergraduate lifestyle and routine.

CHAPTER 3
The Nature of Undergraduate Research

The Structure of Research and Professorships

Before you apply for an undergraduate research experience, it is important that you first understand the basic structure of the research setting. Most students who enroll in university upon graduating high school are unfamiliar with the educational route that a professor followed in order to provide lectures in your courses. Like many lost and confused first year students, I was no different and had to learn along the way. In an effort to fast track your understanding of a professor's credentials, the beginning of this chapter is devoted to sharing what I learned from working alongside numerous professors.

Just like you, once upon a time every professor started as an undergraduate student and completed a Bachelor's degree of some kind. Following their undergraduate studies, those who wish to become a professor pursue at least one terminal degree. A terminal degree is the highest awarded degree in a field of study, and could be either a research or professional degree. In research, the terminal degree is a PhD (or in some fields, an MSc but this is rare), while a terminal professional degree would be a degree that allows one to practise professionally, such as an MD to practise medicine, or a

PharmD to practise pharmacy, etc. While the completion of any terminal research degree could lead to a career in government or industry, and the completion of any terminal professional degree could lead to a career in the practice of the associated profession, those who are interested in a professorship are often required to undergo further education in the form of a post-doctoral fellowship. While this is not always the case, many academic institutions require the completion of a post-doctoral fellowship so that the candidate is highly specialized in their field of study and are capable of planning their own experiments or studies. Finally, candidates who are able to successfully gain a faculty position at an academic institution typically begin at the rank of lecturer, then assistant professor, followed by associate professor, and finally (full) professor. These rankings may vary based on the university and country, however, each educational institution should publish the ranking system they use on their website. Beyond this, some full professors pursue higher positions and become a director of the research institute or become a dean for the faculty.

Finally, understand that all professors who actively conduct research are also required to write grants in order to secure funding for their projects from a government organization, corporation, or foundation (public or private). Often funding is difficult to obtain due to a lack of government contribution or lack of support from private donors, making this process highly competitive. The point I am trying to make here is that every university professor has been studying at an educational institution for at least ten years (usually more), and are extremely hard working individuals with a passion for their field of research and discovery.

As a result, professors tend to choose their students selectively, especially if the position is paid. Often, student salaries (graduate and undergraduate) are paid for through a professor's research grant, unless the student is able to secure his or her own grant. Because of this, if you are an undergraduate it is in your best interest to think about and plan for the type of research experience you wish to gain so that you will be an asset to a professor's research team.

Preparing Yourself for the Application Process

For the remainder of this chapter, I would like to take the time to explain how you can effectively prepare for applying for an undergraduate research experience using a series of strategies that I have found to be highly successful.

Probably the two most important skills you will need to understand before even thinking about applying for a research position are the following: understanding the scientific process and writing effectively.

Understanding the Scientific Process

What I mean by "understanding the scientific process" is having a detailed understanding of the basic process of science as first described by your ninth grade science teacher. This means knowing that every good experimental plan (in the natural and clinical sciences) begins with an introduction, formulation of a question, a hypothesis, a set of experimental methods and materials, a collection of results and analysis of data, and finally a discussion and conclusion of the study based on the findings from the analysis of data. Believe it or not, knowledge of the scientific process is vitally important as an undergraduate researcher, yet so easily and often forgotten or neglected. Of course, while the experiments you will be working on in a real research lab will be much more complex than that of a high school laboratory report, understand that the structure of the scientific process and the order of steps remains exactly the same.

Many undergraduate programs in the natural and clinical sciences offer a course (often required for degree completion) devoted entirely to research methodologies. Having been a teaching assistant for one of these courses at the university where I completed my undergraduate degree, I highly recommend that you enroll yourself in a research methodologies course if you are interested in gaining research experience. This course may be geared more towards research in the natural sciences or the social or clinical sciences, however, the overall goal of these courses are similar if not the same, in the sense that they teach the student how to carry out an experiment from start to finish. Not only will you learn some excellent research skills, but you will also be able to impress any professor who offers you an interview by explaining that you have completed a course in research methodologies and have an excellent grasp of the scientific process.

Writing Effectively

The second skill, in my opinion is much more difficult to master proficiently, and that is, the art of writing. Many science students in university chose their program in part because they disliked writing essays in their high school English or history class, however, the bad news is that in order to successfully complete a degree in the natural or clinical sciences, it is absolutely imperative that you teach yourself to become a good writer.

Unfortunately, the art of essay writing is one of the most poorly taught and overlooked skills in university, where often professors and teaching assistants do not have the adequate time or resources to explain to you how to become a better writer. Instead, they are forced to award you a poor grade on your essay, and your consequent essays unless you make the personal effort required to improve. To make things worse, professors who hire research students often demand that you already possess excellent academic writing skills. If you want to be a good writer, I highly encourage you to be as proactive as possible and take the first step in approaching your instructor or teaching assistant and asking how you can improve the quality of your essay writing. If this is not possible (and there will be times where it is not possible), I would encourage you to ask your friends and classmates who are good writers, attend free university-level writing workshops, and as a last resort hire a writing tutor.

Writing is a skill that takes time and effort on your part, as for me it was a struggle when I first began university. I recommend you do what I did and enroll in a first year philosophy course with the sole purpose of improving your writing and spend the time to ask your teaching assistant many writing-related questions. As strange as this may seem, the subject of philosophy not only teaches you to think critically, but it teaches you to become a proficient writer.

Philosophy, however, will not teach you scientific-writing. Over time as I enrolled in science courses that required a written component, I always made an effort to ask the teaching staff how to improve on my writing skills. I also found it useful to ask for constructive criticism when possible. To a certain extent, I also asked graduate students to look over my written work during my research experiences, as many graduate students are also hired as teaching assistants.

When you are assigned your next essay, write out a draft and edit your own paper multiple times. Read your own writing out loud and remove or re-word any sentences that do not sound quite right, review the quality of the content you have introduced into your paper and ask yourself if what you have written fits the demands of the assignment. Next, check and double check the spelling of words, the grammar of sentences and the structure of each paragraph. Finally, ask others around you to also edit your work provided that they are strong writers. A strong writer once told me that a rule of thumb is that if a sentence or paragraph in your paper does not look or sound right, it probably is not right and needs to be re-written.

The fact of the matter is that improving your writing skills can only ever benefit you. In your final year of the undergraduate degree, many students (are required to) enroll in a senior thesis or project course. This is a long-term research course that often spans the entire eight months of the academic year, where course participants are required to complete a major research project and write a comprehensive experiment report that often exceeds fifty pages, if not more. Even if you choose to not enroll in a senior thesis or project course, learning to write proficiently will greatly improve your grades in any essay you write. Over my undergraduate career, I have evaluated hundreds of student essays as a teaching assistant to multiple courses. From teaching alone, I can tell you that students perform most poorly in writing-based assignments, when compared to other evaluations such as writing tests or giving oral presentations. Thus, if you make the effort to master good writing alone, I guarantee that you will already have a great advantage over the average student in many ways.

Knowledge and Discovery

Identifying the Ideal Research Experience

Once you have mastered these two skills, or even if you are in the process of making an effort to actively make improvements, the next step is to decide on the type of research experience you are interested in pursuing and when you would like to make this goal a reality. Fortunately, there are a variety of different ways to gain a research experience, based on what suits your needs with regards to time commitment. These research opportunities fall into the following broad categories: voluntary, paid or for course credit. When you decide to apply for your first research experience, especially if you are still in your first or second year, you will likely land a volunteer research position before gaining a paid one. Volunteering, while absent of financial benefit, is actually very useful as it provides you with the opportunity to eliminate that catch-22 of requiring experience in order to gain experience. Your primary objective should be to use this volunteering position as relevant experience so as to later gain a paid position in the same research group, or other research groups within the same field.

Voluntary Research

Gaining a volunteer research experience can take place either during the summer or concurrent with school during the academic year. There are benefits and disadvantages to either option, and so you will ultimately have to make the decision as to what you feel is best. In either of the two options, volunteering has one common positive aspect, being that it is not typically as demanding as a paid position or a course credit.

While you should never take your volunteer position lightly, realize that most professors will provide you with increased flexibility when compared to a paid student, since you are not being compensated financially. This may mean that you can choose to volunteer only a few days a week, as opposed to on all weekdays, or that you may be able to come into work later or leave earlier than others. These benefits of volunteering are especially great if you are also taking university classes at the same time during the year or summer school between academic years, since this can prepare you for a paid position during a course-free summer.

One common mistake, however, is that many students feel that these benefits mean that they do not need to work as hard or take their research work as seriously when they are volunteering. Do not adopt this detrimental attitude, because I will tell you before you even apply, that this can only lead to disaster. The reason is very simple. If you become complacent, you risk forfeiting any further benefits you may reap in your research group, including a future paid position. My best advice is to take your volunteer research position very seriously, as supervisors who take on volunteers often look to see whether or not their students possess a strong aptitude for research and commitment to a long-term task.

Be very careful to avoid making overcommitments to volunteer research as you will spread yourself out too thin and likely gain poor results from all of your concurrent activities. Always review your calendar in advance so as to be certain of when you are able to volunteer, and be sure to then discuss the exact details of your research position when you meet your professor. In Chapter 7, I will further discuss how you can adequately prepare yourself for an interview with a research supervisor.

Ideally, if you are beginning your first year as of now, provided that you are settled in and now accustomed to the university lifestyle, you should aim to gain a volunteer position during this time. Though these positions are rare, if you can obtain one, this gives you a head start on the majority of your peers, who likely do not even know that research opportunities exist at the undergraduate level. This will also provide you with a major advantage when you begin applying for a paid position in the summer immediately after your first year.

If you are an upper-year university student and are searching for your very first research experience, I definitely recommend that you consider volunteering, as no matter how far you are in your undergraduate program, many research supervisors only offer paid positions to students holding prior experience. That being said, it is important for you to know that it is never too late to gain experience in research.

Paid Research Positions

The second category of research experiences are the paid positions. These are the positions that I used during my undergraduate years to pay off my school fees, and they are usually best completed during a three or four month summer term. There are different kinds of paid research positions offered to undergraduates, which include co-operative internships (or simply co-ops), work and study (or work-study) positions and summer studentships.

Co-Operative Internships

I briefly mentioned the benefits of acquiring research experience in order to gain a co-operative internship (co-op) in Chapter 2, however, I would like to explain what you can gain from the internship itself. A co-op is effectively a structured method of combining university education with practical work experience, and is designed to help undergraduates make the school-to-work transition. Large and well-established co-op university programs may enroll thousands of co-op students who intern with thousands of co-op employers. The payment associated with completing a co-op term varies widely, and is dependent on many factors including the employer, student's level of education and program of study, however, it is not unheard of that students can successfully pay off a good portion (if not all) of their undergraduate tuition fees through co-ops alone. The main benefit for co-op students, however, is that they are being provided with assistance from their own university or academic department, and are able to gain access to a university website with automated job postings set up by employers. Co-op students are able to upload their resumé, cover letter and job application to this website, and thus be seen by a multitude of potential employers. As a result, some students who successfully secure a co-op position during their time as an undergraduate go on to work for the same employer upon graduation if they demonstrated potential during their internship. The benefit is two-fold because the student is employed quickly, but also because the employer has an incentive to hire a student who already has the necessary training and experience.

Unfortunately, co-op education is not always a positive experience. Often, many students who may not have the competitive edge are excluded by all potential employers, and thus, their graduation date can be delayed by a term if they do not successfully land a co-op job. Additionally, many universities require you as the student to do the majority of the legwork when finding co-op positions, and do not provide as much assistance as many students often believe they provide. As I mentioned in Chapter 2, many co-op programs often require that the student spends more than four years to complete their undergraduate degree because of their co-op program requisites, which was the main reason why I chose not to enroll in a co-op program. A second reason I never enrolled in a co-op program was because I did not want the added pressure of finding a co-op internship at the cost of being delayed a term, as this was a risk I would have to take despite the fact that I had reasonably solid skills with respect to gaining a job. The last reason I chose to opt-out of co-op was because the majority of co-op employers are companies. This meant gaining research experience in industry, as opposed to in academia, which is something I did not want as I wished to gain acceptance to graduate school and pursue my goal of achieving a career at an educational institution as opposed to a for-profit company.

I am by no means discouraging you from enrolling in a co-op program, as there are some great benefits as I mentioned earlier. While a co-op program may be a benefit to one student, it may be a disadvantage to another. As a result, I highly encourage that you understand the risks and rewards associated with making your decision and that you know with certainty why you are making it.

Work-Study Positions

A work-study position is a form of part-time job that is completed concurrently with school. Many universities offer some sort of variation of a work-study position for students, especially in the natural sciences, so as to provide an incentive for students to get involved in undergraduate research. Because of the concurrent nature of these positions, the student is usually not expected to complete copious amounts of work, and may not even be assigned their own project due to the demanding nature of research. This program accommodates for the fact that undergraduates are also studying and completing coursework, and so it is not likely that a work-study job would require more than ten or twelve hours of commitment from the student per week. Though the financial compensation is not the greatest in work-study positions (usually it is no more than minimum wage), gaining a work-study position is a great fit for the undergraduate who has yet to begin his or her career in research.

Like any other research position, there are numerous benefits and disadvantages to work-study positions that you should carefully consider. The greatest benefit of this program is that it consists of a schedule that is flexible around your other academic commitments, though it is important that you make sure to discuss your hours with your research supervisor before you begin. Another benefit is that it is a great way to get your feet wet, when it comes to exploring the world of research, as you will likely be assisting with a graduate student's project or a professor's side project, all while getting paid, albeit, not that much. The work-study position is often dependent on financial need, which is a good or bad thing depending on who you are. If you are a student with financial need this is great as it provides opportunity for you to gain employment while you study, however, if you are not a student with financial need but you lack experience, when it comes to work-study positions you may be disappointed. Due to the fact that you will not be spending a whole lot of time in your research environment during a work-study position, understand that you may not have the opportunity to form as great a relationship with your professor as a student who spends a summer studentship with them. If you want to develop a professional relationship with your work-study professor, you may have to commit extra time beyond what you are being paid to do to get to know him or her better. It is also important to consider that work-study positions may not necessarily provide students with an accurate picture of the research world simply because your commitment to the research, by nature of the program, is too little.

It is important to understand that if you take on your own project, say in the form of a co-op position for example, you should be aware that you will be considerably busier. For this reason, I feel that work-study positions may not provide students with the correct perception of what their work-ethic should be like should they advance further in research. Based on what I mentioned, I feel that work-study positions are great beginner positions, as it will be a great way for you to gain experience and get your foot in the door. They have also helped a few of my friends gain further research experiences within the same research group, such as a research course opportunity or a better paid summer job. That being said, after you complete your work-study experience, if you want to advance further in research, you should ideally gain a research position that requires a greater level of commitment whether it be within the same research group or another one, if you want to gain an accurate understanding of how the research world works.

Independent Summer Research Positions

The final paid research experiences worth mentioning are the summer research positions. These research experiences often span between 12 and 16 weeks in duration, throughout the summer months between typical academic years, and are often best completed during a time period when you do not have other commitments such as summer school. Paid summer research positions range widely in nature.

In the summer after my first year, my professor hired me independent of any departmental program. This is often the case if the professor cannot afford the mandatory minimum stipend required to hire a summer student or if the department simply does not have a summer student program established, allowing the supervisor to assign a student's salary at their discretion. From my experience, these types of summer jobs usually arise when professors have recently acquired their research group or when grant funding is not optimal. It is obvious that the major disadvantage here is that the financial compensation is considerably less than a departmental program, and often the student will work for less than minimum wage since salaries are monthly lump sums and not hourly wages. In this situation, you may also end up having to meet colleagues or collaborators on your own time, since the department will not provide you with any events or opportunities for networking since you were hired independently by your professor. The advantage, however, is the flexibility associated with your summer job.

As a result, you will likely only have to complete any necessary safety training courses and research work that your professor assigns you. This allows you to avoid any departmental red tape that requires you to work for a minimum number of hours a day or weeks per summer, even if you manage to complete your tasks early. As a result, many summer student regulations will not apply to you provided that you discuss this first with your research supervisor.

Departmental Summer Studentships

Some departments allow professors to hire summer students through a department-organized program. These types of paid research positions are known as summer studentships, and are my personal favourite. Summer studentships are a product of departments that have many faculty members looking to hire students, and where grant funding among most professors is usually quite good. In the summer following my second year, I completed a summer studentship with a cancer researcher. The experience is great because since these programs seek students, all hiring professors usually take their summer students more seriously.

Often summer students are required to present their work in the form of a research poster at the end of the summer, and so professors typically assign their student their own research project, which usually makes up a small portion of one of their graduate student's projects. In addition, there are numerous opportunities to network because events are broadcasted throughout the department and sometimes events are created just for summer students. I recall that during my summer studentship, it was mandatory that students attended weekly seminars hosted by different faculty members, which helped students to understand the different types of research being conducted within the department. If you are interested in a summer studentship, understand that your research supervisor will likely expect more of you than if you were hired independently of a departmental program because you are going to be paid more. Also, understand that you will have to adhere to all summer student departmental regulations.

Of course, as you can probably guess, the major downside to summer studentships is their level of competitiveness. As you explore the possibility of a summer studentship further, I am sure I will not be the last to say to you that they are extremely competitive. When I applied for my first summer studentship, I was one of only 50 students selected internationally to join a research group within the department, and all students had to have possessed at least a cumulative average of A minus. My example may be on the more competitive side, but understand that summer studentships are highly sought-after positions. In addition to a standard application form, students applying to a summer studentship are often also required to provide the supervisor or selection committee with a cover letter, curriculum vitae, academic transcripts and even one or two reference letters.

I hope that I have already illustrated to you the importance of gaining reference letters during your undergraduate career, as summer studentships are a prime example of the infamous catch-22 of needing experience in order to gain experience. Not to worry, however, as you are taking the first step to achieving success in landing a research experience by reading this book! In Chapters 5 through 8 I will provide solid strategies that you can employ to prepare to apply for research experiences, based on the summer studentship model, since they are the most competitive.

One final thought worth mentioning regarding summer research positions are the opportunities for receiving undergraduate research grants. While I will discuss this in greater detail towards the end of this book, a short discussion here will help you understand how undergraduate research grants are connected to summer research jobs. Many students beginning university are unaware of research grants and often believe that only graduate students and professors can acquire them, however, undergraduates can too! You can think of a research grant at the undergraduate level as a salary bonus. At the undergraduate level, research grants are awarded based on academic merit and aptitude for research, and serve to encourage the student to pursue research opportunities through increasing their summer salaries. These are great sources of summer income and can be acquired regardless of whether your professor hires you through a departmental summer student program or not. In Chapter 11, I will provide you with further information on strategies for applying for research grants, how undergraduate research grants work, as well as details about why they are important to acquire, especially if your future goal is to pursue a career in research.

Undergraduate Research Courses

The Senior Thesis and Senior Project

The final type of research experience worth mentioning is those that count for an academic course credit. In almost every degree program imaginable there is usually at least one course offered that allows a student to work within a research environment, conduct an experiment, and report on the results of their study. This course is almost always taken during the final year of undergraduate studies, and is known as a senior thesis or senior project. This course is very important if you are planning to pursue a career in research, and some graduate programs even require that you have completed a senior research course. Senior theses or projects are worth anywhere between a one half-year course credit up to a two full-year course credit, though the typical research course is worth one or one and one half course credits.

As a result of the project's length, unless you have worked actively in a research environment for more than a year, a senior thesis course will likely be the longest and most comprehensive research experience you can possibly gain during your undergraduate degree. In fact, you can almost think of a senior thesis as a shortened Master's degree, since you will likely have a supervisory committee (i.e. more than one research supervisor), need to write a major research report, and often defend your thesis or present your findings at an end-of-year undergraduate thesis symposium. Unfortunately, for many students this is their very first research experience, as there is typically no prerequisite of having gained research experience in order to enroll in this course. As a result, many students struggle and feel that the commitment required for a thesis course alone is extremely overwhelming.

Apart from the research tasks themselves, if you enroll in a thesis course you should be prepared to commit more time per credit than your average textbook and lecture-based course. For example a typical one half-year course in biology may require three hours of lecture attendance, one hour of tutorial class and two hours of laboratory time per week, which equals a grand total of six hours of commitment per week per half-year course taken, excluding preparation and study time. This means that if your senior research course is worth three half-year courses and spans over one academic year of two terms, you should be prepared to spend more than 18 hours on research work alone each week, excluding preparation time. When I completed my senior thesis in the Department of Medicine, I recall that on average most of my weeks consisted of 20-25 hours of research, though I had some friends who would spend upwards of 30 hours per week on just their senior thesis alone. Sometimes prior research experience, however, can reduce the time you will need to spend on your project per week, simply because you are a more efficient researcher. Based on my observations, students who had gained experience in research prior to their senior research course managed the challenges and obstacles of their research project considerably better than those who jumped into it without ever getting their feet wet.

Preparing for Your Senior Research Course

Besides having prior research experience, there are a few other skills you should work on in order to excel in your senior research course. The first skill is learning to write very well. I have emphasised this earlier in this chapter, and can only continue to overemphasize the importance of this skill. My final undergraduate senior thesis consisted of more than eighty pages, including figures and references. The written component of my thesis alone consisted of fifty pages. While this may be lengthier than the average undergraduate senior thesis, be prepared to write exceptionally and excessively if you enroll in a senior research course. As I mentioned before, learning to write both efficiently and effectively at the academic level will greatly speed up this process while simultaneously reducing your stress associated with writing such a large paper.

The second skill you should familiarize yourself with is a good understanding of research statistics. While not all projects will require the same amounts of statistical analysis, and you may not necessarily be responsible for computing all the statistics associated with your research project, at the very least you will need to be able to interpret the statistical analyses associated with your research results. A good start would be to take a course in introductory statistics, which is usually offered at the first or second year of undergraduate science programs. Take good notes in this class, and retain your textbooks. When I was a teaching assistant for a research methodologies course, part of my tasks included teaching students how to interpret and compute simple statistics. Before I taught my lesson I recall asking my students whether they had taken a course in statistics, and by the end of the class it was clear to me that those who had experience in statistics performed better. Unfortunately, statistics is not a mandatory prerequisite for some undergraduate programs in the sciences, however, if you want to excel in research, especially if you plan on undertaking a senior thesis or even graduate school, knowledge of basic statistics will definitely be useful.

A final skill that will help you prepare for your senior research course is excellent oral communication skills. In research, being able to present your findings is just as important as conducting experiments. While you may not necessarily have to make a presentation to your supervisor during a summer research job, I can almost guarantee that you will have to present your findings orally at some point during your senior research course. During my senior thesis, I presented my results orally three times, twice at two separate research symposiums and once in front of my supervisor. Before orally presenting your thesis, practise your public speaking skills in front of your friends, and instruct them to ask you challenging questions about your project so that you are certain you have a solid understanding of your project.

Better yet, if you gain a volunteer research experience in your first or second year, ask your supervisor if you can have the opportunity to present your project in the form of a poster or presentation in front of your research group. This not only helps you to prepare for a day when you will be evaluated for the quality of your oral presentation, but this will also show your supervisor that you are taking the initiative to practise your oral communication skills.

If you can master these three skills and gain a research experience or two before beginning your final undergraduate year, I have full confidence that you will be able to complete your senior research course or project more efficiently and effectively than the vast majority of your peers because you will have a good grasp of what research is about.

Research Placement and Practicum Courses

Another way to prepare for your senior thesis or project is by completing a research placement or practicum course a year before you enroll in your senior research course. These types of courses are gaining popularity at many universities, as academic institutions attempt to provide students with greater assistance in gaining real-world experience and senior thesis preparation. During my time as an undergraduate I really appreciated that my university offered these courses, which I took complete advantage of in order to broaden my experiences in research. In a sense, these courses can be thought of as smaller senior research courses, and usually do not require as much time commitment as a senior thesis. If you are beginning your third year and have not gained any undergraduate research experience yet, these courses are especially beneficial for you, as they will help you know what to expect during your senior research course. When I enrolled myself in these courses, I was able to practise my skills in research, academic writing, statistics and oral communication, things that I would end up doing again during my senior thesis, except while completing a considerably larger project.

Applying Your Knowledge of Different Research Experiences

Take advantage of the fact that you now have a very good idea of all the types of research experiences offered at the undergraduate level and you understand how to adequately prepare for them. Using this knowledge, you will have a significant advantage over your peers when you seek out the different research opportunities that exist at your own university. If you are just beginning your undergraduate degree, do not worry too much about finding that perfect research project as it is quite often the case that students at this level do not even know what career field they want to pursue. At this stage, just focus on gaining an experience that you enjoy, and set reasonable goals for yourself regarding what you would like to achieve during your upper years and future summer terms.

I encourage you to treat your undergraduate research career as a step-by-step process allowing you to better understand how the research world works with every experience you gain. You should think of your undergraduate studies as a time for experimentation in research, during which you can practically join any research group that is of interest to you, provided you have the necessary course-based skills to conduct research in that field. I encourage you to take full advantage of this benefit and privilege.

CHAPTER 4
Searching for a Supervisor and Research Group

Identifying Your Research Interests

Now that you have a good understanding of the types of research experiences you can gain during your undergraduate career, it is now time to explain how to prepare for your first research experience and what you can expect once you begin. Recall back to Chapter 1 where I mentioned that the opportunities offered in research are very broad. You may need to take some time to think about what field interests you the most, while keeping you post-undergraduate goals in mind in order to best tailor your research experiences to your future endeavours. A common misconception among undergraduates is that all research is conducted in a laboratory. I am sure you have read enough of this book to know that this is not true, but perhaps you may still be uncertain of what your tasks or duties could include as an undergraduate researcher. Research takes place in every field of study, including those outside of science, however, I will provide you with some examples most relevant to the student in the natural and/or clinical sciences.

A good way to start thinking about what research field would interest you is by browsing through your university textbooks. Your first-year biology textbook, for example, is likely divided into sections based on sub-fields such as microbiology, physiology, evolution and genetics, etc. Similarly, you could do this with any textbook in chemistry, physics or the environmental sciences, based on the subject that interests you the most. Once you have found a sub-field that interests you, visit your university's faculty web pages and see if anyone conducts similar research.

A quick look at different faculty web pages at your university should indicate to you that research programs vary widely. Most life sciences or physical sciences students begin their research career in the natural sciences, and I was no different. I began my undergraduate research career in a wet-laboratory as it was what was familiar to me from my biology textbooks. If you choose to join a wet-laboratory, as the name suggests, you could be working with a variety of chemical or biological materials, varying from chemical reagents, to bacteria or fungi to human or animal cell cultures and tissues. Though projects vary widely, if you pursue a research experience in a molecular biology laboratory, there are numerous transferable skills you can learn should you pursue this type of work in the future. For example, you may learn how to operate a microscope, quantify proteins using assays, perform DNA electrophoresis, perform polymerase chain reactions, or grow and culture cell lines or bacterial cultures, as examples. If your experiments require animal models, you may work with lab mice or rats, and learn how to handle and anesthetize rodents, perform injections, and complete surgeries.

If you are not interested in working in a wet-laboratory, but still want to gain research experience in the natural sciences, consider other options. Many of my classmates found great interest in field work. If you enjoy working outdoors with animals, opportunities exist to work on conservation biology projects and investigating animal behaviour. If you are a student who is more mathematically-inclined consider research work in bioinformatics or biostatistics. Students who wish to stay out of a wet-laboratory can gain an experience in a dry-laboratory working with computer models that simulate anything from chemical reactions to weather conditions to animal populations in the ecosystem.

Many health sciences students along with students who are interested in a future in healthcare may also express an interest in gaining a research opportunity in the clinical sciences, as opposed to the natural sciences. Clinical science is a branch of medical science that determines the effectiveness and safety of medical therapies. Again, the research opportunities available in this field vary widely, and it is indeed possible to gain a research experience that is both wet-laboratory based and clinical in nature, as some research groups may incorporate techniques used in both the natural and clinical sciences. Beyond this, it is possible to also gain a clinical research opportunity outside of a laboratory setting. For example, as part of the experimental protocol, you may have the opportunity to interview or survey patients enrolled in a clinical study, develop patient databases or gain knowledge of healthcare policies.

Obviously, it would be impossible for me to provide you with a complete list of skills you could gain from each and every research opportunity, but I have provided you with some of the most common examples. Regardless of what interests you, once you have decided upon a field of interest, remember that if you are trying to gain your first research experience, you should make it a point to keep your options open. This is especially true if you are in your first year and are not quite sure of what types of research opportunities exist within your field of interest.

Funding is Not Allocated Equally

Following my first year, I did not even know what academic research meant, and so choosing a field that interested me was difficult. I ended up emailing more than 300 professors appointed to either a life sciences or physical sciences department, and only received two summer job offers. I gratefully accepted my first job offer to work in a biogeochemistry laboratory where I would spend the next three months investigating the effects of heavy metals on cyanobacteria, however, knowing what I know today, I would not have applied to my first research experience in the same fashion. While applying for my first research position, I wish I had known where most of the research funding was being allocated and who was acquiring this funding.

Understand that the majority of research funding allocated to researchers is supplied by governmental organizations, and that there is a heavy bias in providing greater funding within certain research fields, while less in others. If I think back to the 300 professors I applied to during my first year, the majority of them were researchers in physical chemistry and developmental, evolutionary or conservational biology. Though it should have been obvious to me, generally speaking, it is most often the case that researchers addressing the most publicized issues of our society today gain the most funding. This means working alongside researchers who study highly-prevalent, chronic diseases (i.e. cancer, diabetes, heart disease, HIV/AIDS, etc.), address important and pressing world issues (i.e. global warming, mental health disorders, drug and alcohol abuse, etc.) and contribute to fields that have made recent and significant advances (i.e. stem-cell biology, regenerative medicine, nanotechnology, etc.). I have provided just a few examples of the most well-funded research fields, however, if you wish to know more about all the well-funded fields for yourself, visiting a governmental research institute web page should provide you with all the answers you need.

Select Well-Funded Research Groups

Be aware that universities that possess greater levels of research intensity are always more well-funded. This usually means that the larger the university, the greater the number of research programs and research professors available, and the greater the opportunity for gaining an undergraduate research experience in a well-funded research group.

The reality is that you will gain a richer research experience with a better funded research group. During the summer after my second year I gained a summer studentship at a very well-funded cancer research laboratory, and the difference between my first and second research positions were vastly different. As an undergraduate, understand that you only have a very limited amount of time to complete your research project and so the more project-related work you can complete, the more you will be able to learn and accomplish.

Well-funded research groups allow you to bypass the tedious non-research tasks, such as creating stock solutions or washing glassware and pipette tips. In fact, some extremely well-funded research professors may even hire a laboratory technician or research assistant that is paid to complete these non-research tasks so that the research staff can complete their tasks more efficiently.

I am by no means encouraging you to neglect cleaning up after yourself, or demand that someone else complete all your non-research tasks associated with your project, as it is very likely that you be responsible for all this at one point during your job. My main message is that by working in a well-funded research groups, there is greater potential to work more efficiently.

Determining how well a research group is funded is a task in itself that you must commit to completing before gaining a position. There are a few ways of figuring this out without directly approaching the professor leading the research group in question. One way, of course, is to evaluate whether their field is one of the well-funded fields as I explained earlier, however, this just provides a generalization. If you get the opportunity, meet with a graduate student or another undergraduate student who is already working within the research group, and ask them what the research environment is like. Also, look at the professor's faculty web page. While many students do not even browse a faculty web page before meeting the professor, doing so allows you to acquire a goldmine of information. Many of my peers, even during the final year of my undergraduate year, never trained themselves to assess the little things that make big differences on a faculty web page. As a student, you can think of a professor's faculty web page as a professional Facebook® page where he or she showcases all the highlights of their research group and academic career. Beyond just the professor's academic degrees and research interests, the next time you are on a faculty web page, look out for the following: academic ranking of the professor, size of research group (and number of graduate and undergraduate research students), number of laboratory technicians/research assistants, and number of published research papers (published per year).

Understanding How a Supervisor's Academic Rank Affects Your Research Experience

In Chapter 3, I explained that teaching faculty begin their careers at the rank of lecturer, followed by assistant professor, then associate professor, and finally (full) professor. To an undergraduate this may seem irrelevant because if you do not plan on becoming an academic faculty member at an educational institution you may not have even noticed that a faculty ranking exists all together. The reality, however, is that the rank of your supervisor does matter, and sometimes it matters quite a bit.

You likely would not be able to gain a research experience with a lecturer, as lecturers are hired by the university with the sole purpose of teaching. They may work for a research group, but it is probably not the case that they are leading one as a principal investigator. Assistant professors, associate professors and professors, however, can all be considered as principal investigators to research groups. This means that they have the authority to supervise undergraduate and graduate students, provided that they have the funds necessary. Working for a supervisor of any given rank comes with both benefits and disadvantages, and so it will be up to you to decide which suits you best.

Due to the fact that assistant professors are newly appointed professors, they usually do not have the greatest amount of funding. In fact, some assistant professors have no funding themselves, and only collaborate with other faculty members on research projects, and so working for someone like this may be difficult. Many assistant professors are often also required to teach university courses, and may not be able to gain as much protected time as they wish to conduct their research. Another thing to be aware of is that some competitive professional or graduate programs view reference letters from assistant professors as not as important as a more distinguished faculty member, though this is not always the case. That being said, there are also multiple benefits associated with working for an assistant professor. For one thing, they are usually able to provide more time to guide their students for two reasons. The first is that they are typically younger, motivated and working towards a higher ranking, and are thus often at their office, as opposed to travelling abroad. The second is that they naturally have a smaller research group because of a lack of funding, and therefore, spend more time on average with each student than a faculty member who leads a large research group.

The size of a research group is a very important factor that you should consider as an undergraduate. A small research group is beneficial as it will most likely allow you to spend more time with your supervisor, and also allow you to form closer relationships with your colleagues. In contrast, professors leading larger research groups usually have considerably more funding, and are thus able to employ additional people, make more discoveries and publish results faster. Joining a small-sized research group was the strategy I chose for all of my research experiences during my undergraduate with the exception of one. Oddly enough, my least favourite research experience was that one exception because I felt I could never form a long-term relationship with any of the laboratory members nor did I have enough time to receive feedback from my professor regarding my performance. Not everyone feels the same way, however, as I knew many friends who enjoyed working within a research group with more people. As a result, you may have to join multiple research groups over your undergraduate career to determine what size works best for you.

You should also be aware of how many graduate and undergraduate students are employed by your professor. Graduate students are typically highly knowledgeable about the skills sets required to conduct research within their research group, hence, it is best if there are a few that will be around to show you the ropes when you are first hired.

Other undergraduates, however, especially if hired at the same time as you, may not be as knowledgeable. When you begin the research position I would advise that you avoid a research group that is led by a professor who hires multiple undergraduate students simultaneously. While this may sound a little strange (after all, undergraduates are your peers), if you want to succeed in completing your research project, your most productive environment will be one where you can ask others for help who are more knowledgeable than you. Furthermore, if you join a research group with five or ten other undergraduates, it is highly unlikely that your professor will be able to form a long-term meaningful relationship with just you (unless you put in major effort), and thus, you risk working for an employer who might forget your name before you even approach him or her for a letter of reference.

Finally, be aware of whether there is a laboratory technician or research assistant who is a member of the research group. A general rule of thumb is that the more non-student research assistants that are hired by the principal investigator, the more well-funded the research group.

It is also important to consider other titles and credentials that a potential supervisor may possess. For example, if a professor is a (junior) scientist they likely will have less protected time for research than a senior scientist. Senior scientists are often usually provided with this title because their research group has been highly successful and productive.

If a professor is an associate dean or a dean of the university department that means that they have additional educational commitments (i.e. managing graduate applications, developing curriculum) and may not devote as much of their time to their research program, while if a professor is a director of an institute, they may spend extra time advocating for their institute or travelling, as opposed to spending as much time with the students in their research group.

Finally, if a potential supervisor holds a professional degree they may also hold down a secondary job. For example, a professor with an MD or PharmD in addition to their PhD may also be a practicing healthcare provider, and thus would not spend all of their working days in a research-based environment.

It is very important to look out for these things, as well as to ask potential supervisors what their research commitments are like if they are able to offer you a position, so that you are not unpleasantly surprised by their agenda when you begin working with them. If you take all of these factors into consideration, you can effectively eliminate the difficulty of finding the right balance of positives when selecting a research group to work with.

Assessing the Publication History of a Potential Supervisor

Now that you understand the importance of assessing a professor's academic ranking and research group size, it is time to look at their academic publications, the last key piece of information found on a professor's faculty web page. This is the most direct line of evidence for a research group's productivity. Usually, most faculty web pages will have a section that displays a professor's research history. If not, by making a quick visit to a free academic article database such as PubMed.gov or Google Scholar™ and simply entering the researcher's full name, you should achieve the same results.

While this may not provide a perfectly complete picture of your professor's life work in academia, it should provide enough information for you to evaluate the productivity of his or her research group. First, count how many research articles show up when you search a professor's name, but be careful, as common names can lead to hits for papers not published by the professor of interest to you. This number can range from less than five papers, if the professor recently completed their PhD, to a few hundred if the professor has been an exceptionally productive researcher over the past few decades.

Next, count how many papers your professor has published over the last five years or so, as this is a good indicator of recent funding and productivity within the research group. If one or two papers are being published annually, you can assume that funding and productivity is adequate, while if the same number of papers are being published over a span of three or four years, then this may be a cause for concern. Having worked for professors on both ends of the spectrum, from one publishing less than one paper per year to another publishing over ten papers annually, I can tell you that this information matters to you as a student.

Be aware that the number of papers published annually by your professor should only be used as evidence of productivity and funding, and not certainty of a research group's present situation. For example, a professor with few publications over the recent past may have been ill or unable to work for that period of time. Professors who are clinician-scientists also fulfill a professional job, and as a result, may not allocate as much time to their research program. Alternatively, a professor may have recently made an important discovery that has not yet been published, or may have received a major research grant capable of supporting many research students this summer. In contrast, a professor who has published many papers in the past year may have coincidentally completed many different multi-year projects during the same year. Lastly, it is important for you to take into account the length and quality of a professor's research publications. Not all publications are equal, and some journals are more competitive to publish in than others. Sometimes professors from prestigious institutes may simply publish less material because they choose to only publish the results of long-term, high-quality studies in the best academic journals.

Though there are many factors to consider, what you should aim to look for is a sense of consistency. Evaluate for yourself whether the publication rate of the research group in question is similar each year, increasing or decreasing. You will then be able to decide whether the research group in question is productive and worthwhile your time.

Understanding Authorship Number

One final point of importance is evaluating the professor's authorship number on recent publications. While this may sound too technical, any researcher will know that there is a major difference between a first author and last author on a publication.

Before I explain why, you need to understand what the authorship number actually means, as this was not even something I was aware of when I was beginning my first year of university. In the world of research, authors on publications are listed out based on the quantity of their contribution to the research paper. This means that the researcher who completed the most work on the research project (regardless of academic rank or credentials) will gain the position of first author, the researcher who completed the second-highest amount of work on the research project will gain the position of second author, and so on and so forth. The last two or three authors are usually reserved for the supervising professors who contributed considerably less to the project, but provided the majority of the funding.

Obviously, a research group can be identified as productive if you can find the names of research group members on the published papers, however, also look to see where the research group's professor name is listed. If the professor's name is very last, this likely meant that that professor funded the majority of the project, and was also the main person to acquire the funding for the study. Instead, if you see that the professor's name is listed as the second or third last author on a publication, along with many author names who are not members of the professor's research group, then this likely means that the professor gained authorship via collaborating with another research group with similar research interests. This means that the professor likely did not contribute major funding towards the project.

Keep this information in mind, as this will allow you to determine whether your professor is funding the majority of his or her projects or not. Also, be aware of publications that are review articles. These are articles that do not qualify as primary research articles, and are a summary of primary research articles written on the same topic. Research articles written by professors or graduate students do not require the research group to spend any funding on experimentation.

Realize that the process of evaluating all the highlights listed on a research professor's faculty web page does not take very long to complete. As a result, I would highly recommend that you review all of these factors, after narrowing down the list of professors you are most interested in working for, as this will help you prepare in the event that they invite you for an interview.

Reminders for Before You Apply

You are now very close to beginning the actual application process. There are just a couple more important things that you should know before you actually begin applying for a research experience.

The first is a reminder to keep your research interests broad, especially if this is the first research experience you are applying for or if you are at the beginning of your undergraduate degree and are unsure of your career aspirations. As I mentioned earlier in this book, if you are preparing for your very first research experience you are already at a disadvantage when compared to those students who even hold one prior research experience. Furthermore, if you are in your first year, you should be aware that your competitors include upper year students who have taken more advanced courses and who have a stronger understanding of specialized topics. This being said, the vast majority of research techniques are often learned by the student on the job. For this reason, I even know a few students who completed their first research experience towards the end of their high school years, even though they did not necessarily have a strong university-level science background. Because the odds are against you as a first year student who lacks research experience, you must make up for it by preparing more thoroughly for the application process than your competition. This is the most important piece of information you should know, so always keep it in your mind throughout every step of the application from the planning stage up until the moment that you are offered a research position.

The second is to make a schedule for yourself as to when you would like to achieve your goals, while also taking into consideration what research positions you would be open to accepting before you begin this process. This will help you to avoid making a poor decision when you are already applying for positions and meeting with professors, and when you may not necessarily have as much time to make important decisions.

Considering Non-Research Factors

Taking into account all the non-research factors of your research experience can be rather tricky, and there are often instances when one of these factors becomes more important than a research factor. By non-research factors, I mean the location of the research position, other commitments you will be responsible for while concurrently taking on undergraduate research, and when the research experience takes place, among other examples that do not have anything to do specifically with the research itself.

It is important that you consider the location of your research positions, so that you are aware of where you might be living or commuting. For many first year students, moving away from home for the first time can already be a stressful and emotionally-difficult journey, and so you will have to decide for yourself whether making a move to another unfamiliar location away from social supports is something that you can manage in a positive and productive way.

Beyond living away from home or your home university, understand that commuting is another option. It is important, however, to decide whether this is worthwhile both with respect to your time and finances. For example, during one of my summer research jobs, I recall that I commuted one and a half hours each day to the research lab, and another one and a half hours back home at the end of the day. Looking back, if I had wanted to increase my productivity, three hours on the road each day may not have been the best decision. Financially, however, it was considerably cheaper for me to live at a location that was more affordable (i.e. with family, away from the downtown core, etc.) and make the long commute each day. Also, consider the nature of the job. Fortunately, the money I spent commuting was only a small portion of my research job salary, and so I did not make any financial loss, however, if you are offered a voluntary research experience, you will need to consider the cost of your travel expenses to get to a location where you will be working without financial compensation.

It is usually best that you take advantage of volunteer research positions when you lack experience, as you will likely make the money you needed to commute back through a future research job. It is better to lose a little money in order to gain the experience than to forgo the volunteer position entirely and risk becoming a less competitive candidate for a research job the next year. I would definitely not advise, however, that you spend your entire undergraduate career volunteering if you have the adequate experience to acquire a paid position instead.

In the event that you are completing an undergraduate research course (i.e. research placement, practicum, project or thesis), I definitely recommend that you choose a supervisor who works at the same university campus as where you take your undergraduate classes, when possible. I was careful to ensure that both my senior thesis supervisor and co-supervisor had their offices and research groups at the campus where I attended my undergraduate classes, so as to ensure that I could meet with them immediately if necessary, and complete my thesis work at the same location where I was taking my courses. Remember that undergraduate research courses are already demanding in nature, requiring many hours of commitment per week. Based on the experiences of my peers who chose to complete their research courses with professors at off-site locations (i.e. hospitals on the other side of the city), I can tell you that by the end of their research experience they had deeply regretted their decision.

Managing Your Research Commitments

Though I briefly mentioned this earlier, I want to reinforce the importance of evaluating your commitments to other activities when considering a research position. This is especially important during the school year, as you will be busy with coursework, among other commitments, such as a part-time job or sports team. Before you start applying for a research position, make a list of the things you will need to do, and then make another list of things you want to do, during the academic year. After scheduling in all your mandatory activities, decide how much time you have left to commit to research and any other optional endeavours.

If you are a proactive university student you should already be well aware of how much you can manage during a stressful academic year, as you want to ensure that you do not overcommit yourself to too many activities. When I was an undergraduate, I found that a really effective strategy was to aim to add one extra activity to my agenda each academic year, provided I could maintain my grade point average. If I realized that I could not maintain my grades, I would consequently drop the extra activity from my schedule. This strategy allows you to test your mental, physical and emotional limits, helping you to learn the difference between a challenging agenda and an overwhelming one. A challenging agenda is when you are able to successfully balance all the items on your agenda (i.e. school, research, student group commitment, sports team, etc.) and feel positive about yourself, while an overwhelming agenda is one where you feel unable to successfully make a 100% commitment to each task, and are burned out from the process. It is important that you understand that what appears to be an overwhelming agenda at the beginning of your undergraduate career may be only half as challenging in your final year. Understand that your perceptions, capabilities and abilities change as you grow as an undergraduate, and so if you cannot manage a research position in your first year, this does not mean that you can make the same conclusions about your second, third or final year.

When your research experiences take place also makes a big difference based on your level of undergraduate education. I highly recommend that you make a long-term, rough timeline for your research goals during your undergraduate degree. This may seem difficult, but it really is not if you start simple, and add the details in later. For example, before your begin your research career, you may have a goal of gaining a volunteer research position in first year, summer studentships after both your second and third academic years, and a senior project course in your fourth (final) year. From here, you can then begin to add in the details as they become clearer to you. Perhaps after your first year, you decided that you really enjoy the field of cancer biology and want to be an oncologist. By knowing this, you can then focus on spending your summer studentships at cancer research laboratories, and then completing your senior project with an oncologist who studies the clinical outcomes of chemotherapeutic drugs. You may also decide along the way that you would like to co-author a paper for a cancer research journal, or apply for an undergraduate research grant that funds undergraduate cancer research. As you can begin to see, by gaining research experience you also gain the benefit of acquiring real-life practical experience that can truly shape your future path, even if you are not quite sure of your career goals and aspirations just yet!

I sincerely hope that by now you are not only encouraged to gain an experience in research, but you are confident in preparing to gain one. With this in mind, in the next chapter of this book I will provide you with detailed strategies you can use to successfully apply for your first undergraduate research experience.

CHAPTER 5
Constructing Your Application

Preparing Your Research Application Documents

You are now about to begin your application process, one of the most crucial steps in gaining research experience. What you do (or not do) during this process will determine whether you gain a research opportunity, and if so, the quality of that experience. I have devoted this entire chapter to providing you with the best research experience application strategies that I have acquired and perfected with each research position I gained during my undergraduate career.

Before you contact a potential supervisor expressing your interest in their research or register your name on a summer studentship online application portal, there are certain documents that you should prepare first, so that you do not need to rush to complete them later. Upon hearing from you, most professors who are interested in hiring a student will request one or more of the following five items: a cover letter, your curriculum vitae (or resumé), a letter (or contact) of reference, a sample (or two) of your academic writing, and a copy of your academic transcripts. Notice that none of these items are easy to acquire at a moment's notice, and thus, it is essential for you to prepare these documents ahead of time so that you will have all five items ready to send off immediately following a potential supervisor's request.

The Cover Letter

Be Aware of What You Convey Indirectly

The most important item on the aforementioned list is the cover letter as it is what all supervisors read first. Past experience has taught me that if your cover letter is poorly written or contains content a supervisor dislikes, you can safely assume that the rest of your application will be going through the paper shredder unread. While it is imperative that you ensure that your five-item application package is without flaw, particular emphasis should be placed on ensuring that your cover letter is nothing short of perfect.

A cover letter tells a potential supervisor many things, including messages that undergraduates may not even realize they are sharing with their supervisor. While your objective is to concisely share your skill sets and past experience with your supervisor, you are also indirectly sharing your writing ability, your enthusiasm and interest, and your ability to communicate concisely and effectively. If you do not understand, allow me to explain. In other words, while many students are too busy focusing on sharing why they are a great candidate, they fail to realize how they are actually choosing to conveying this information.

Your writing ability is reflected through the spelling, grammar and sentence structure of your entire cover letter. Any trivial errors will give your supervisor the idea that you have not taken the time to edit and proofread your cover letter before sending it. Not only does that appear careless, it appears lazy, and I assure you that both of these characteristics are not appreciated in the research setting. Allow me to provide a personal example to demonstrate how easy it is to make a major mistake. Back in my first year, after I had written out my letter and checked it over a couple times, I figured it was written well enough and I began emailing it out to professors. Due to inadequate proofreading, I had not realized that I had written the word "contentious" instead of "conscientious". While this was a minor typographical error, while one word has a very positive meaning, the other has a very negative one. After emailing a few potential supervisors, one professor kindly brought this to my attention by stating that he thought I meant "conscientious", but also apologized in advance in case I had truly meant "contentious"! As a result of this experience, I now always proofread my writing multiple times over, and I highly recommend that you do too if you wish to avoid a humiliating result similar to mine. If you want to write a stellar cover letter, be sure to write and edit multiple drafts, and try your best to get friends or colleagues with good writing skills to proofread it. Better yet, if you know any upper year undergraduates or graduate students, ask them if they would be willing to help proofread your letter.

Your enthusiasm and interest is reflected through your tone of voice throughout the reference letter. Throughout my time as a teaching assistant to a research methodologies course, I have read through my fair share of cover letters, and unfortunately, too many exemplify dull writing. Strive to create a cover letter that delivers a message in an enthusiastic and positive fashion. By this I mean that you write confidently about your skill sets, use descriptive and specific adjectives in your sentences, ensure that you tell your potential supervisor that he or she is free to address any further questions or concerns that they may have regarding your application, and finally, remember to thank the professor for taking the time to read your letter.

Finally, your ability to communicate concisely and effectively is reflected through both the length and content of your cover letter. No cover letter should exceed more than one page, especially at the undergraduate level. Any longer than that and most potential supervisors likely would not have the time or interest to read that much text. Be careful that you avoid a crucial mistake that many students at the undergraduate level seem to make regarding content. That is, include only what is important for your research supervisor to know, and not what is important to you. This is a research application, not a university program's supplementary application, and so while well-roundedness may be appreciated in the latter, it most often is not valued in the former. It is highly unimportant to divulge information about your sports team, non-research related student organization, or non-research part-time job involvements, unless they specifically reflect an acquired skill that is directly transferable in the research setting. Such skills include critical thinking, the ability to work effectively in a team, and dedication and commitment to the task at hand. What is important is information about your academic accomplishments and training, past experiences you have gained in research (at the undergraduate level, even if in an unrelated field), past experiences in teaching, and any general research-related skills you possess. My best advice to you is that if you cannot relate one of your non-research related opportunities to one of these four categories of information you should include in your cover letter, then simply do not mention it. To further clarify, allow me to explain why these four items, in particular, are vitally important to your cover letter.

Academic Accomplishments

Your academic accomplishments and training is probably the most important piece of information you can share with a potential supervisor if you do not possess any prior research experience, as this is the only way that they can honestly assess you as a candidate. While anything else that you make a claim to is only based on the opinions of you and/or your referees, your academic accomplishments can be clearly verified by your academic transcripts.

I told you earlier that you do not necessarily need good grades to gain a research experience, and I will now explain why. Like most students in their first year, I performed poorly on a few of my courses. In fact, I would never score such low grades again for the remainder of my undergraduate career. If this situation sounds familiar, then like me, you will need to employ the necessary strategies to avoid the attention that may be given to your poor grades. In your cover letter, draw particular attention to your most relevant and highest scored grades. For example, after my first year, I highlighted in my cover letter that I had scored high grades in biology and chemistry, the two fields in which I was considering gaining a research experience. Another strategy is to use the word "range" (i.e. A-range grades) as you can then highlight a subject without drawing particular attention to the grade itself. If you acquired any grades below a B-range, I would advise that you do not mention the grade or the subject at any point in your cover letter. Instead, be mentally prepared to explain why you achieved this grade during the interview process, which I will elaborate on in Chapter 8. While you should highlight certain subject grades, do not mention anything about your transcripts throughout the entirety of your cover letter. By far, this is one of the most common mistakes I have seen students make, as unless your entire transcript is superbly flawless, you are effectively providing your potential supervisor with an offer to view the courses in which you performed poorly. When it comes to transcripts, my advice is to avoid any mention of this document all together, while being prepared to produce a copy if and only if you are asked to do so.

Finally, I will provide you with a strong word of caution since many supervisors require that you produce an unofficial transcript. This is often done so as to help the student eliminate the time and cost of requesting an official transcript, however, this also allows for the opportunity to fabricate your grades. Not only is this punishable as an offence of academic dishonesty, but if your research supervisor finds out after you are hired you can expect your research opportunity to be terminated immediately. Consider that your professor may also divulge this information with his or her colleagues, so that you are permanently blacklisted from any future opportunities in the department of your preferred field of study. This risk is never worthwhile considering that there are many professors who are willing to provide students with the chance to conduct research within their group even if they are not the strongest academically.

Past Experiences in Research

The second piece of information that you should include within your cover letter is any past experiences you have gained in research. This section will vary widely depending on your undergraduate year and becomes increasingly more relevant as you gain more research experience in related fields. For example, when I applied for a research position with a stem-cell researcher, I explained that I had been previously employed at a molecular biology laboratory, and possessed relevant research skills in cell culturing and protein extraction and quantification, which helped me to acquire the research experience. Even if you have held a research experience in the past unrelated to a research position for which you are currently applying, it is worthwhile to mention it. It is better to declare that you have experience in an unrelated research field, than to allow a potential supervisor to believe that you have no research experience at all. After all, even if you will not be using any specific transferable research-methods, you are still indicating to your potential supervisor that you have a good understanding of the scientific process and the nature of research in general.

If you do not have prior research experience and this is the first position you are applying for, you will have to be more creative when completing this part of your cover letter. This is your opportunity to show your potential supervisor that you have the ability to confidently showcase your skills despite not having held any prior research positions. During the first year of university, any undergraduate enrolled in the natural sciences should have enrolled in at least one or two courses with laboratory components, such as an introductory science course in biology, chemistry, physics, etc. As a result of taking such courses, you can then explain in your cover letter that you have gained specific laboratory-based skills or, when possible, even state that you scored especially high grades on the laboratory components of your courses. The same strategy can be applied to students in the clinical sciences, who may have enrolled in course with a research component such as introductory anatomy or epidemiology.

Teaching Experience

Next, you should include a section on your past experiences in teaching. This can be anything from mentoring elementary school students in sports, to peer tutoring high school students in mathematics, to serving as a teaching assistant to a university professor, though you should only list the most recent and relevant experiences. Including a section on teaching in your cover letter indicates a few positive things about yourself. For one, it shows leadership and critical thinking skills, as you are explaining to a potential supervisor that you are able to help transfer the knowledge you have gained to someone who has not yet learned this knowledge and that you are able to answer a student's questions that you may not have initially expected. Being able to transfer the knowledge you have as a researcher, even at the undergraduate level, is extremely important if you plan on writing an abstract or presenting a research poster of your project.

Sharing with a potential supervisor that you have past teaching experience also shows that you have taken on roles in the past that demonstrate your trustworthiness. No one would ask a dishonest student to take on the role of teaching another student, nor would a professor want a student conducting research dishonestly.

Finally, indicating that you have served as a tutor (particularly a university teaching assistant), shows your potential supervisor that your academic abilities are stellar. If you apply for a teaching assistantship for a certain course, you are required to have scored extremely well in related courses (including the course for which you are applying) and/or possess great experience with the subject matter of the course, through some other means. Interestingly, I was hired as a teaching assistant to a course in research methodologies due to the fact that I had extensive research experience, however, I am certain that regardless of which experience you gain first, you can use it to help yourself in gaining the other.

Research-Related Skills

Finally, the last piece of information you should include in your cover letter is any general research-related skills you possess. These are skills relating to your personality, work ethic and general academic abilities. While you may have alluded to the majority of these skills when writing about the first three categories of information in your cover letter, you may want to reiterate your strengths by drawing out certain skill sets. This could mean stating that you have excellent academic writing abilities, oral communication skills, patience, leadership, trustworthiness, dedication, etc. Ideally, you should provide evidence of an activity that you have partaken in to demonstrate the validity of each characteristic you use to describe yourself. Just be sure to only highlight the most important skills, however, otherwise it will just appear as if you have carelessly listed all the positive adjectives you could think of to distinguish your abilities.

The Cover Letter: Final Reminders

Remember that a cover letter is a written piece that takes time to create. If you do not take the proper care to write it, then it would be unfair that you should expect a potential supervisor to read it. If you include all four pieces of information about your academic accomplishments and training, past experiences you have gained in research, past experiences in teaching, and your general research-related skills, in a well-balanced fashion, I have full confidence that you will be able to write an attractive and success-inducing cover letter. If you would like to view an example of a cover letter that you can refer to as a model when applying for a future research experience, please see Section 1 of Appendix A.

The Resumé and Curriculum Vitae

Understanding the Difference between a Resumé and a Curriculum Vitae

Once you are satisfied with your cover letter, you should then turn your attention to constructing your resumé or curriculum vitae. Many students just beginning their undergraduate (and even some well into their upper years) either do not know the difference between a resumé and curriculum vitae, or believe that the terms are interchangeable.

Before I explain how to construct your own resumé or curriculum vitae for your research position applications, it is important that you know where, why and how the one document is distinct from the other. While outside Canada and the United States, a resumé and curriculum vitae are often used interchangeably by employers (i.e. Europe), in North America, the main difference between a resumé and a curriculum vitae is the length of the document. While a resumé aims to highlight your most notable achievements and accomplishments in as few words as possible, a curriculum vitae is designed to provide the reader with a highly comprehensive document that contains all of a person's accomplishments throughout their entire professional career. Furthermore, a curriculum vitae is written for a very academically-oriented audience, while this is not necessarily the case for a resumé. This means that if you are applying for a job in the non-profits, government or the private sector, you should provide your potential employer with a resumé, while if you are applying for a job in academia (i.e. professor, clinician-scientist, researcher, graduate student etc.) you should provide your curriculum vitae.

The resumé and curriculum vitae are similar in the sense that they should both contain a professional summary of qualifications. This is a summary statement of approximately 100 words that attempts to capture your best qualities. Your summary statement should be a brief list of your professional qualifications, written in approximately five lines. As an undergraduate, a good summary statement would include your undergraduate year, program and degree, your research or employment background, teaching background, a sentence stating why you are ideal for the research position, and some work-ethic skills that best describe you.

Both documents also contain the following sections: contact information, employment, education, awards, extracurricular and volunteer experiences and relevant skills and (non-academic) training. Provision of this aforementioned list of items results in a fully complete resumé, but only a mere starting point for a curriculum vitae. Often, even though these items are included in both documents, the curriculum vitae contains far more detail. In addition to these items, a curriculum vitae would also include the following sections: relevant course work, academic research experiences, research grants received, a description of your theses (i.e. for a senior thesis course or a graduate degree), academic papers or books published (or in review), abstracts, posters or talks presented at academic conferences, and teaching history. In general, while a resumé should be no more than two pages, regardless of the nature of the job you are applying to, the curriculum vitae of a distinguished professor may well exceed 50 pages or more.

Making the Transition from Resumé to Curriculum Vitae

As you can see, if you begin gaining undergraduate research experience early, it can be easy for you to acquire many of the items that belong on a curriculum vitae. While you will not gain 50 pages worth of experiences throughout your undergraduate degree alone, it is possible that you will acquire more accomplishments than what can be fitted onto two pages alone. If you have recently graduated from high school, you likely do not have prior experiences in research, and for the most part only ever needed a resumé. This is because of the nature of the jobs you applied to, for whether you worked for a summer camp, restaurant, retail store or newspaper delivery service, your employers would only ever ask for your resumé at most. Now that you have taken the first step towards achieving an undergraduate degree, it is time for you to evaluate how you can use your existing resumé (if you have one) as a template to construct a curriculum vitae that will suit the many needs of positions available to you in higher education.

Before my undergraduate career, I had a small and undetailed resumé similar to most high school students. The beginning of my undergraduate degree marked a time where I updated my resumé, tailoring it so that it was more appropriate for applying for research positions and university-level extracurricular or volunteering activities. Because I continued to pursue research opportunities throughout my undergraduate degree, I began to construct what some call a "hybrid resumé-curriculum vitae". This is an ideal balance between a resumé and curriculum vitae as it acts as a starting point for a student's fully-fledged curriculum vitae. Finally, towards the end of my undergraduate career and as I applied to graduate schools, I created a fully-fledged curriculum vitae. This means that my curriculum vitae was (and still is) written in a format that is reflective of a standard academic curriculum vitae.

Two Important Pieces of Advice

The most important piece of advice I can provide you with is to not underestimate the time it will take you to even create an elementary resumé. If you have never written a resumé or your resumé has been untouched for a year or more, you should ensure that it is fully completed and updated before you apply for a research position. Understand that the majority of time you spend on writing your resumé is not necessarily typing it up on a word processor, but instead, finding documentation of your accomplishments or even spending time trying to remember some of your past accomplishments.

My second most important piece of advice is to never list any item on your resumé or curriculum vitae that cannot be proven through a record or reference letter. For one, this obviously means that you should not lie on your resumé, but secondly, this also means that you should not declare any item that you cannot prove you completed or accomplished. I say this for a good reason, because if you cannot prove your involvement in a task even though it was completed honestly, it will look as if you lied on your resumé. To avoid this situation in the future, always ensure that you acquire some form of documentation that you had a part-time job or completed an extracurricular activity. Some items on your resumé such as your education or scholarships can easily be verified by the academic institution, so you do not need any documentation in this case, however, for everything else, documentation is a necessity.

Develop a habit of asking your boss, supervisor or coordinator for a brief record or reference letter that explicitly states how long and how often you made commitments to the task in question, immediately following your completion of an activity. If you have not done this for past activities, and if it is not too late, go and ask for the records or reference letters to activities you have not acquired documentation for as soon as possible. Towards the end of my undergraduate career I knew of numerous friends who struggled to verify that they completed certain tasks on their resumés, yet all this could have been easily avoided if they had simply ensured to request documentation immediately after completing each activity. Keep all these documentations in an organized binder and retain it until the day you graduate or until you no longer need them. Not only will this help you to write your resumé and curriculum vitae, but it will also help you when you apply to graduate or professional schools that ask you to declare your time commitments to each activity you have participated in over the past years.

Constructing Your Resumé: What to Include

If you are just beginning your resumé, start by including all your contact information and your personal summary of qualifications. Next, consider what you wish to include in each section of your resumé. In academia, it is highly recommended that within each section you add each entry in a chronological order, beginning with your most recent items at the top. This allows your potential supervisor to quickly assess what you have accomplished most recently.

As a general rule, it is unnecessary for you to include any accomplishments on your resumé that you have acquired before you began high school. Not only is this irrelevant and unimportant to your potential employer, but it is also a waste of time and valuable resumé space for you.

As this resumé will be used for applying to an undergraduate research position, your first section should be your education. Information that you should include is as follows: your university, the program you are enrolled in (i.e. life sciences), your undergraduate year, the year you enrolled and your expected graduation year.

Some students also choose to include their grade point average, however, only do so if you believe it will provide you with an advantage. Remember that a resumé is supposed to be written with the purpose of impressing a potential employer and expressing to them that you are more competitive than your competitors, therefore, if you declare that you have a poor, average, or even slightly above average grade point average, your potential supervisors may immediately become disappointed with your resumé.

At the undergraduate level, you should also include details about your high school, especially if you graduated from an academically enriched program or graduated with stellar grades. Keep in mind that as you eventually become an upper-year undergraduate, even some of your high-school accomplishments will be of very little value on your resumé.

Your next section should be your employment history. While I advised that you do not include jobs unrelated to research in your cover letter, as a recent high school graduate it is understandable that you will have worked in a non-research-related field and so your resumé is the place to declare this. If you had indeed held a part-time job concurrently with your high school studies, it is important that you are able to convey such commitments to a potential supervisor. Under each job, provide a brief description of your tasks. Try to include characteristics that demonstrate that you would be an asset in a research environment. For example, if you worked in retail, you could express that you were able to work diligently as part of a team or handled stressful situations well, since both these characteristics are very much needed if you wish to perform as an undergraduate researcher.

Next, create a section for awards if you have acquired any thus far. For research position applications, focus on including awards of an academic nature. Declaring that you received an entrance scholarship to your undergraduate program, for example, is far more valuable in research, than having gained an award in a sport or hobby. I am not saying that you should not list a prestigious sports award, as at the beginning of your undergraduate degree it demonstrates commitment and dedication to a task, but as you continue your undergraduate degree and begin acquiring more academic awards while revising your curriculum vitae, you should be prepared to remove less academically relevant awards and accomplishments. Understand that the world of research is highly academic itself, and thus, many achievements outside of academia are not particularly valued by some faculty within the field.

Your next task should be to create a section that lists your most important and relevant extracurricular and volunteer experiences. Again, when describing these items, try to include characteristics that demonstrate that your abilities will allow you to thrive in research.

The last itemized section of your resumé should be your relevant skills. This is the least important section of your resumé, so do not make this section too lengthy. Begin by listing the most important skills you have, such as command of more than one language, knowledge of computer spreadsheet programs, and experience with laboratory safety and the scientific method, then list the less important work-ethic skills, such as critical thinking skills, enthusiasm or dedication, all of which cannot necessarily be verified simply by looking at your resumé or interviewing you alone.

Lastly, many students in the past have asked me about the inclusion of a references section. This is a section that I advise you to write with extreme caution, if you so choose to do so. I personally include nothing but one line that reads "References available upon request." in my references section of my resumé. Understand that potential employers may take advantage of the fact that you listed your references and their contact information. For example, they may call them before they consider you for an interview. This is an especially bad situation if you have not yet told your references that you are applying for the position, which could result in a reference leaving a poor impression of you on a potential supervisor. It is also difficult to alert all your references of incoming calls or emails if you are applying to dozens of professors at one time. A smarter method would be to just declare that you have references available, and provide your employer with their contact information upon request. That way you can have control over this situation, and can be the first one to tell your references that you are applying for the research position instead.

Constructing Your Curriculum Vitae: What to Include

Once you have acquired and completed your first research experience, you are most likely ready to begin constructing your hybrid resumé-curriculum vitae. Save a copy of your resumé as well, so that you have the two documents handy as you will likely use both during your time as an undergraduate student.

In your curriculum vitae, you should begin including all your research positions you have held to date, and put them in a section above employment. Always ensure that the most important items are listed first, while maintaining the rule of ordering your items chronologically. Sometimes this requires some creativity, however, and this is another major reason why constructing a curriculum vitae can take more time than you would expect. I was perhaps an untraditional high school student, having never held a part-time job before enrolling myself in university. It turns out that my first research experience was also my first part-time job, and so I decided to morph my research and employment sections together. A strategy like this allowed me to include all my paid and unpaid research experiences under the same section, and prevented me from having to place research-related volunteer work in the less merited volunteer section. If you do wish to retain some of your non-research related employment history, consider creating a research section, followed by an employment section, then finally a volunteer section. By doing this, you can now include all your research experiences (paid and voluntary) in your research section, thus making it easier for potential supervisors to see your entire undergraduate research history.

As you progress in undergraduate research and begin presenting research posters or acquiring research awards, be sure to include these in your curriculum vitae under the appropriate sections as well. Another important aspect of a curriculum vitae includes the provision additional details under each item. While in a resumé you may simply list a job (research or non-research), teaching appointment, volunteer experience, or extracurricular activity you have held with a one line description, your curriculum vitae should include a few lines devoted to explaining in greater description, the nature of the job, and your positive contributions and skills learned. Awards should also be described in your curriculum vitae as well, and you should include the date you were awarded, the reason for being awarded, and if applicable, the monetary value of the award. Abstracts, posters and oral talks that you have presented at scientific conferences can also contain a description as well, though this is not necessary. Research positions, both paid and voluntary, should also include a brief description of your research supervisor's credentials, academic rank and your research project, since these details may be important for a future potential research supervisor or graduate and professional school application. If your goal is to attend professional school, you may also want to include a section on job shadowing, as this can show that you are committed to learning about the career you are interested in by spending time observing the nature of the job. This is also a section that is attractive to researchers, as they will see that you are dedicated to gaining greater experience by approaching your career goals from both the research and clinical perspectives.

Update Your Resumé and Curriculum Vitae Frequently

My last words of advice for you regarding both your resumé and curriculum vitae are to update both documents regularly. It is far less overwhelming to add a new item to either document shortly after you have completed or achieved it, than to add numerous items right before you need to provide a potential employer with your resumé and curriculum vitae. Making regular updates are also advantageous because you will avoid the possibility of forgetting to add an item or any of its associated details. Personally, I usually add in new items the day I receive notice that they are confirmed. In addition, I update my resumé and curriculum vitae at the very beginning of every month, and have been doing this since the beginning of my second year. I also put a line in the header that states the month and year of my update so that both my employer and I can quickly identify when the document was last updated. During each monthly update I review my two documents for changes that have been made over the past month. This means, for example, that I add end dates to jobs or activities I have since left over the past month, or introduce additional details that I was not yet aware of at the time I received confirmation of a research project or award.

The Resumé and Curriculum Vitae: Final Remarks

If you have been following along so far, you will realize that both the cover letter and the resumé or curriculum vitae are all documents that require time and care to create and maintain. You should also have realized that they all should not be constructed or updated at the last moment based on a potential employer's request, but instead should be completed well in advance of beginning the research application process. The good news, however, is that if you have completed each of these documents, you are nearly ready to begin sending out your applications. If you would like to view a sample hybrid resumé-curriculum vitae that you can use as a model when applying for a future research experience, please see Section 2 of Appendix A.

The Reference Letter

Earlier I mentioned that you should acquire a reference letter for each of the items on your resumé and curriculum vitae that cannot be verified by records, in order to confirm your time commitments. When applying to a research position, however, some potential supervisors may request a reference letter or contact that can comment on your academic abilities, research ability, personality and work-ethic. In this case, the aforementioned reference letter that simply confirms your time commitments is not enough and you will instead need to approach a previous employer, supervisor, or professor who knows you well enough to comment on the information that your potential employer is specifically asking about.

When you make contact with a potential research supervisor for the first time, do not explicitly state or imply that you can provide a letter or contact of reference. Instead, let them ask you if they need one. I have gained some of my undergraduate research experiences without having the need to submit a reference letter, and so if you can avoid this step, it is by all means to your benefit. If by chance you are asked for a reference, be sure to confirm whether your potential employer is asking for an entire reference letter or simply the contact information so that they can personally get in touch with your referee on their own time. Either way you want to ensure that your referee is someone who not only knows you well, but who is someone that you can depend on to speak positively about you. If your potential supervisor simply asks for the contact information of a referee, the process is relatively simple.

It is always best to have a few referees in mind, so that you can choose who you think would be best for the position for which you are applying. It is also always a good idea to ask your referee for his or her permission before providing a potential employer with their contact information, and it is also important to alert them before or very shortly after you are made aware by your potential employer that they will be getting in touch with your referee. Either way, you should choose a referee who understands the nature of academic research, or better yet, who is a part of it. A reference for your first research position may be someone who is not a researcher, such as a high school teacher, teaching assistant or better yet a course lecturer, and this is completely understandable, however, as you progress in research and begin applying for a competitive summer studentship or an undergraduate research award, it is usually required that your referees are academic faculty involved in research. If you apply to graduate school, it will definitely be the case that all the referees who write your letters must be professors who are also principal investigators.

Why You Should Practise Reference Letter Writing

It is also worthwhile that you practise writing your own reference letters about yourself, as this will help you in a few ways. By acting as a referee to yourself, you will realize how to promote your strengths and understand what types of words and sentences look attractive to a potential employer. Writing your own reference letter can also help you to write a better cover letter or curriculum vitae as you will be able to read these documents from the perspective of a hiring supervisor more easily. One final reason is in case your referee does not have the time or ability to write your reference letter. Although this is not ideal, it is important that you are prepared for this in the event that it happens. Unfortunately, sometimes professors have too many tasks at hand when you need a reference letter, and thus only agree to submit one if you write a skeleton draft first. When circumstances do not permit you to ask another referee or if you do not have another referee that you can ask, it is better to have your referee to sign their signature on a reference letter that you have written than to not have a letter at all. Realize that submitting a reference letter that you wrote should be your last resort, and if you wish to avoid this situation from happening frequently, remind yourself to always ask your referee for a letter one to two months before your application is due so that it is written and received well in advance of the deadline. If you are looking for an example of a self-written reference letter, I have included one I wrote in my third year in Section 3 of Appendix A.

The Writing Sample

Of all the five research application documents, a sample of your academic writing is the least often requested, however, in the rare event that you are requested to produce this document, it is always a good idea for you to be prepared in advance. For this reason, you should never delete the academic assignments and essays that you have written for a high school or university course. Instead, keep these word processor files in an organized folder on your computer, and back them up on an external hard drive periodically. Producing a sample of your academic writing abilities can be the easiest task if you retain these assignments and essays, yet it can be the hardest task if you had already carelessly deleted them from your computer.

It is also important that you keep graded copies of your written work (electronic and printed) so that you can remind yourself how well you scored in your assignments. This way, if you wish to submit a writing sample, you can look at your evaluator's comments and edit your assignment before sending it off to a potential research supervisor.

If you are asked for a sample of your writing, always ask about what type of writing your potential supervisor wants of you. Some professors may ask for a research essay, just to evaluate your writing skills, while others may ask for a research abstract, poster or even grant application, so that they can evaluate your writing skills within the context of research. Either way, it is always ideal to provide them with a written piece that is closely related to their field of research. For example, if a professor of stem cell research wishes to see what your academic writing abilities are like, it would be better to send him or her a copy of an essay you wrote in a university-level cell biology course than an English course.

Academic Transcripts

Finally, the last document you will likely be including in your research application package is your academic transcripts. Since you have virtually no control over the contents of this document, I have little to say about this final document. Be aware, however, that some universities charge a fee to print transcripts, so always ask a potential supervisor if you can send an unofficial copy (i.e. a photocopy) or an electronic copy of a your degree audit instead. Also, understand that many universities make it a policy that they are provided with a week or two of time between the day that you place the order and the day that it arrives at your requested destination. Though there is usually an option for you to rush an order, it is still important that you plan ahead to ensure that this portion of your application is delivered on time.

The Completed Application Package

Once you have acquired your cover letter, resumé or curriculum vitae, letters of reference, a sample of your academic writing and a copy of your academic transcripts, this completes all the documents you will likely ever need for a research position application, with the sole exception of the position-specific research application forms themselves. By preparing these five important documents in advance, you can be assured that you will be able to produce these documents in a far timelier manner than your less prepared peers. Remember that the nature of research is competitive, and so staying one step ahead of your peers is crucial, especially if you have not yet acquired any prior research experience.

CHAPTER 6
Applying

Overview of the Application Process

This chapter covers everything you need to know about applying for an undergraduate research experience.

The preparation of applications is not only a challenging process both in and out of research, it is also stressful, tedious and competitive. Whether you choose to pursue a career with the government, the private sector or even academia, the key to getting the job is through a good application. If you are applying to professional or graduate school, you can be certain that the difference between students who are accepted versus those who are rejected is the quality of their applications.

In research, applications provide that first impression to the recruiting supervisor, and thus, in order to even be considered for an interview, you must be able to show that you are a candidate who is nothing but stellar. If you are well-prepared, the stress associated with the demanding nature of applying for undergraduate research experiences can be largely reduced. For example, by following everything mentioned in the last chapter of this book, you will have all the documents you need to impress your potential supervisors before you even approach them. Many of your peers who begin applying for a research experience for their first time likely will not be aware that a potential supervisor will ask for a cover letter or their curriculum vitae, and as a result, they will likely complete these documents in a rushed and careless manner. Considering that you are prepared, you can use any additional time to complete any research application forms and carefully prepare for your upcoming interviews instead.

Avoiding the Common Application Mistakes

If you know what I know about undergraduate research, you will be surprised to find out that the vast majority of students approach the application process in the wrong way. As attractive as it may be to begin your research career by applying to the most advertised undergraduate summer studentships at your university, I highly advise against this idea. If you are a student at the beginning of your undergraduate degree and possess little or no prior research experience, more often than not this method will result in frustration and failure. The reason, of course, is quite simple as you will soon see.

Learn to Think for Yourself

During my own undergraduate career, I learned that many students do not tend to think for themselves, for reasons that I do not understand. Whether it is taking a course rumoured to be easy, or seeking inadequate advice in choosing a degree program, many undergraduates tend to lack the responsibility needed to make decisions for reasons that are personally sound. Gaining research experience because your peers are getting involved or because it is a prerequisite to something else are both poor motives for conducting research, though I have heard many of my peers provide these same reasons. If you are one of these students, I urge you to reconsider this type of mindset as experience has taught me that these are the same students who generally lack direction and leadership. While I would consider that I have always been one to think for myself and make each of my decisions with good reason, my undergraduate experiences reinforced why doing so is crucially important.

Knowledge and Discovery

Understand that when you think for yourself, you are the one in control. You are proactive, not reactive, and you will be skeptical enough to realize when not to follow the crowd and adopt their mistakes. One of the best examples of this, as you may guess, is applying for undergraduate research positions. Learn to examine why most students who wish to gain research experience apply to the most advertised undergraduate summer studentships, and assess the consequences of this behaviour. If you use this approach during every step of applying for a research position, I promise you that things will become a lot clearer.

It is only logical that you should conclude that students apply to well-advertised research positions because it does not require any further work of searching for an experience. When your high school teachers and university professors said that you will no longer be fed by a silver spoon in university, they seriously meant it. Back in high school, if your science teacher told the class of an opportunity, if you readily applied for it, you would have most likely succeeded, whether it was membership in a student organization, a volunteer experience or an additional academic-enrichment opportunity. In the high school setting, this approach worked because the teachers put students' interests first, the opportunities were less competitive, and the number of competitors was far fewer. The undergraduate world is vastly different, and the sooner you realize this, the easier it will be for you to understand that you need to make drastic changes to your methods of acquiring opportunities. In university, this same high school approach will not work for 99% of students, because if a professor openly announces that an undergraduate research position is available, he or she will be looking for only the most stellar students, the opportunity will be extremely competitive, and the number of competitors will include more students than you will ever have the ability to meet in your entire undergraduate career.

I recall that one of my first year instructors made an announcement in class one day, explaining that ten stellar students who submitted an application to his email address would receive a volunteer research opportunity lasting no more than a week. This was not a bad idea on my professor's part, as it helped him to gain some volunteers for his research program very quickly, however, from a first year student's perspective, the benefit of applying is little if not absent all together. As I expected, the applications poured in immediately and all ten volunteers were selected before next day's class, leaving the vast majority of applicants disappointed.

If you want to avoid a situation such as this, it is imperative that you use a more proactive and creative method of applying for your future research positions. For these reasons alone, despite having gained numerous research opportunities throughout my undergraduate career, I never once applied to a research position that was advertised in a lecture hall or on an online application portal.

How to Find Research Positions

The following strategies are neither complicated nor difficult. In fact, any student willing to put in the effort with the right mindset can implement them, yet they are powerful because the vast majority of students are not aware that research positions can be obtained in this fashion. As a result of perfecting these strategies over my four years as an undergraduate, I am now able to provide you with the information you need to successfully (and painlessly) gain a research experience.

As a general rule, if you wish to achieve the best results, always plan to apply early. The right time to apply will differ from position to position, but if you are able to contact professors one academic term away from when you want to join their research group, this usually provides those hiring with enough time to consider your application, assess their research group's space and funding and decide whether you are the right fit. To put this in perspective, this means that you should strive to begin your application process four to five months prior to the month in which you wish to begin your position, regardless of whether you plan to apply for a position during the summer or academic year.

If you have made a decision regarding what research field interests you, as I explained in Chapter 4, then begin compiling all the names and email addresses of the professors who lead research groups that are of interest to you. There is no specific number that you require, however, if you are applying four to five months prior to a research position, fifty contacts should be sufficient even if you have no prior research experience. I highly recommend that you also keep a second list of contacts in addition to these fifty, just in case you have limited success. Understand that this number will change based on many factors. For example, if the nature of the research position is voluntary, more professors will have available positions and if you apply too late as I did in my first year, you may experience limited success, and may need to apply to more professors. Finally, if you have prior experiences in research, specifically within the same field, your number of offers will increase drastically. In my third year, I applied to ten fairly well-distinguished professors, and received three offers, a stark contrast to having applied to three hundred professors in my first year, and receiving only two offers. It is clear to me that the reason I received more offers with less applications was because I was better prepared, carried prior research experience and applied early.

Methods of Applying

While there are a variety of different ways in which you can apply for a research experience, not all methods are equally effective.

Inquiring about the prospects of a research position through phoning unexpectedly or showing up unannounced at a professor's office door are probably two of the most difficult methods, as you need to be well-rehearsed and aware of what you should and should not say. For this reason, I do not recommend this approach unless you possess flawless communication skills and have a stellar ability to address any unexpected questions and concerns.

Instead, some students request to book an appointment with the professor, either through the professor's secretary (if he or she has one) or through the professor himself. While this demonstrates your interest to a potential supervisor, it is a fairly inefficient method of applying as you may be making an appointment with professors who already know that they cannot afford to hire another student. I also advise against this strategy.

Other students ask graduate students or teaching assistants that are already a part of the research group they are interested in joining. This is not a bad idea, as gaining a position through connecting with your network is always something positive, however, at the undergraduate level it is not always easy to make connections, especially if you are still in your first year. This is a good approach to use when you are in your senior undergraduate years and have already built a solid professional network.

Finally, there are those students who communicate with professors via email. I have found that by using this method, I have gained far greater success than using any of the other methods combined, as it is a quick and efficient way to get in contact with multiple potential supervisors all at once. It is for this reason that I recommend this approach, as I will explain later in this chapter.

Keeping Your Online Record Clean

Before explaining why I recommend the use of email to apply for research positions, let me first provide you with a few words of caution regarding the email address and account that you use to make contact. Understand plain and simply that a professor is not your friend, nor should you consider him or her to be your friend. A professor is a working professional, and should be considered as your employer. As a result, you should evaluate whether the email address you are using is appropriate for use in a professional environment. Your email address should be clean and simple (i.e. firstname.lastname@email.com), and should never contain anything that could be deemed even remotely inappropriate. Better yet, use your university assigned email address if you have been provided with one, as this will provide any potential supervisor that you email with proof that you are in fact a university student. Finally, ensure that your email address is not linked to any inappropriate websites or non-professional social networks.

Realize that it is not uncommon for employers, including professors, to search your name on popular social networks in order to gain (sometimes unwanted) first impressions of you. A simple Google™ search of your name or your email address should allow you to see what potential supervisors can view of you, and if you find anything that you would feel embarrassed to discuss further at a job interview you should take the necessary steps to remove it.

Either way, once you have a clean record online, take advantage of emailing, as this is often the fastest and most effective method of getting in contact with a potential supervisor.

The Email Advantage

I highly recommend emailing because it provides you with numerous benefits. Through email you can prepare a well-written message at your own pace and generally replicate certain sentences or paragraphs when contacting multiple professors, all of which cannot be done over the phone or in person. It is important that you send off all your emails within the same period of time, so that you can determine how many interview offers you will receive before meeting with any potential supervisors in person. This means that you should strive to write your emails out for each of the professors you are applying to first, and then send them all out within the same business weekday.

You may feel very overwhelmed by the fact that you will need to send out at least fifty emails, however, you will soon see that this process is really quite simple if followed in a stepwise fashion. When you first contact a professor by email, you do not know how likely it is that they will accept you, nor do they know your aptitude for research. As a result, it is not necessary for you to specifically highlight what interests you about their particular research program in detail, especially if you have no prior research experience. Remember that if a professor is interested in hiring a student, he or she will want to evaluate your performance on an interview as opposed to your knowledge of their research program through an introductory email. Instead, focus on writing one generic email that sounds highly personal, which you can then serve as a template. You can then make alterations to this draft when personalizing emails being sent to different potential supervisors.

The following are some example phrases that you could use which sound personal but are written generically:

- "I am asking whether I could join your laboratory as an undergraduate summer student, as I am very interested in your research program."
- "Throughout the year I enjoy attending various academic colloquia relating to your field of research when time permits."
- "I am very interested in pursuing a career in [professor's field], hence, I am emailing to ask whether you would be able to provide me with the opportunity to gain some research experience under your supervision."

If you read carefully, you would notice that all three of these sentences have one thing in common. If you received this email as a professor it would seem highly personal, however, as a student, these same exact sentences could be used in your email to address multiple professors. This is precisely what I mean by a generic email that sounds highly personal, allowing you to make a positive impression on multiple potential supervisors rather quickly.

That being said, I would not recommend that you send off completely identical emails to each professor. Instead, create small personalization details in your email so it appears as if you did not send a completely generic message. Always address a potential supervisor by their title and full name (i.e. Dr. John Smith), and never by "Sir", "Madam", or worse yet, "To Whom It May Concern", considering you are sending a message to a personal email address.

You should also acknowledge your potential supervisor's department or institute in a subtle manner, such as by stating the following: "I understand that you are a researcher at the [educational institute and department] and conduct research within the field of [professor's research field], which is of great interest to me." Also, always be sure to include information regarding how the professor can contact you and when you are available to meet him or her.

Finally, always write your email in a tone that showcases your optimism, enthusiasm and confidence, as this can go a long way when it comes to impressing a potential supervisor. Remember that each professor has travelled a lengthy and difficult road to investigate and discover in a field that is highly interesting to them, so they will only be motivated to hire students who find their work to be equally interesting.

You can think of this email as a shortened cover letter, which I explained how to write in detail in the previous chapter. The reason I say that it is shortened, is because you want to keep in mind that many professors are very busy, and as a result, read their emails quickly. If you make it a page long, you are likely providing too many details that are unimportant to your potential supervisor. If you can accurately summarize your cover letter in half a page, while retaining all the important information mentioned above, you should be on the right track. If you are still unsure, I have provided you with an example of an effective research application email in Section 4 of Appendix A, which you can refer to when beginning your own application process.

Addressing a Lack of Replies

As you will likely find out, sometimes professors will not reply to your emails. In my first year when I applied, I would estimate that I never received a reply to roughly 40% of the emails that I sent out. While it is often the case that many professors do not have a research position available, or are uninterested in hiring an undergraduate, some may genuinely have neglected your email or allowed it to sink to the bottom of their inbox and become forgotten. You can think of emailing as analogous to a telemarketer who is cold-calling customers. While a few supervisors may be in need of a student, the majority who do not either turn down your offer or ignore your request.

If you feel that you have not received enough (or any) offers one week following the day you sent off your emails, then do not hesitate to re-send your emails again. This strategy has often helped me, and I have even been thanked for my persistence and determination! Keep in mind, however, that there is a fine balance between persistence and aggression, so it would not be wise to send any more repeat emails if do not receive a reply one week following your second email. You do not want to risk angering a professor by spamming their inbox.

Another strategy that has helped me has been to phone a potential supervisor, or better yet, show up to their office and try to meet them in person. As I mentioned earlier, this is a more difficult strategy, and so I would only recommend that you do this with a select few supervisors who you are most interested in working with, and only after adequate interview preparation. This strategy is trickier because while you will need to be ready to communicate that you would like to gain a research experience under a potential researcher's supervision, you must also be prepared to answer any questions that they may have. In the next chapter, I will provide you with strategies that you can use to help prepare yourself for the interview process.

Responding Efficiently and Effectively

Provided that you contacted a few dozen potential supervisors, within a couple hours of sending out all of your emails, you should expect replies from at least some of the professors you had contacted. For the rest of that day, and throughout the following week, it is important that you check your email's inbox vigilantly so that you can respond to the emails you receive in a timely manner. If a professor replies stating that they are unable to accommodate your request, it is only polite that you email them back thanking them for their time. Many students fail to do this and make a mistake by doing so, as this could leave the professor with a negative impression of you. After all, they may one day become a future colleague or collaborator. Furthermore, when you reply, you should also consider asking whether the professor knows of any colleagues or collaborators who are currently looking for an undergraduate research student to supervise.

The most important type of email you will receive, however, is when a professor either wishes to discuss your request further or invites you for an interview. If this is the case, you should be as diligent as possible in confirming a date, time and location to meet. Most professors are extremely busy and so their days are often packed with meetings or appointments from the moment they step into their office to the time when they leave, so schedule an appointment to meet with him or her as soon as possible. Also, be sure to address any questions, concerns or requests that the professor has brought to your attention in their reply. This may mean clarifying why you are interested in their research group, why you are an ideal research student, or simply providing a writing sample. Either way, do not ignore any questions or concerns, but instead make sure to address everything the professor inquires about in your reply or explain that you would like to discuss them in person. In your reply, you should also remember to ask whether you should bring anything to the meeting or interview. In addition to whatever they suggest, and regardless of whether they suggest you bring anything all together, always ensure that you pack a copy of your curriculum vitae, cover letter, writing samples and academic transcripts just in case. It is better that you end up not submitting these documents, than realizing that you need it once you begin your interview. Finally, remember to thank the professor for his or her time and for providing you with an interview opportunity, but do not thank them for a research position, unless they have explicitly stated that they can guarantee you a position. Even then, it is usually best that those words are saved for when you meet your professor in person.

Understand that communication at this stage of the application process is extremely crucial to your success. If you can word your thoughts carefully through email, and articulate your ideas and answers to the professor's questions thoroughly during the interview, you might just be able to avoid any discussion regarding your weaknesses, such as poor grades on an academic transcript. This is precisely the key to gaining undergraduate research experience even if you do not possess stellar grades. I will return back to this idea in the next chapter.

Avoid Accepting Offers Online

Finally, a few words of caution, as some professors sometimes provide you with an offer through email. Many undergraduates, especially in times of desperation, make the mistake of accepting an offer immediately. I recommend that you resist the impulse to do so as this can cause a multitude of problems. If you accept the offer through email, you risk giving up the opportunity to have a (more) thoughtful discussion in person, or even worse, you may simply just be asked by your professor to show up to the research office on the first day. Because the application process is largely completed online, some professors may not even feel the need to interview you in person.

Based on my experiences, you should always meet a professor in person before accepting a research position. If you do not do this, you may find out later that you lack an interest in your research project, your professor does not put in enough effort to supervise you, or you do not get along well with your supervisor's research group members. All this could have been simply avoided by asking good questions during your first meeting with the professor. Furthermore, if you accept an offer online, many students do not take into account that they may receive future offers. As a result, you may be forced to choose between forfeiting a better future offer or ruining the relationship with your current professor by turning his or her offer down only a few days after accepting it.

Completing Research Application Forms

Before I move on to the next chapter it is important that you understand how to complete site-specific research application forms successfully. If you have been following along, you know that these site-specific research application forms have been mentioned before. These application forms are usually associated with competitive summer studentships offered at well-known departments within a university. These application forms contain additional questions for candidates to answer, beyond the typical five-item application package I mentioned in the last chapter.

Knowledge and Discovery

Perhaps I am jumping a little ahead, as you may end up completing these after the interview process, however, there will also be times when it may be advantageous for you to complete these forms ahead of time, such as if you are required to complete these forms before contacting research professors within the department.

Read and Ask Questions Carefully

Before you begin completing any application form, always be sure to read through the entire document as well as any instructions or frequently asked questions on the department's website. If there are still components of the application that you do not understand, read through the department's website and the application form once again to ensure that you did not miss anything. If you still cannot find any information to clarify the component of the application that confuses you, only then should you email the administrative coordinator responsible for the research program. Regardless, it is always best to find out about information you do not understand than to reduce your chances of acceptance to a summer studentship by answering a question incorrectly on your application.

Understanding and Assessing Your Eligibility

Before beginning your application, it is also important for you to understand the eligibility requirements associated with the summer studentship. Usually the two main factors that determine eligibility are your education level and your grade point average. Other factors that may be weighted as important (but are not necessarily used to determine eligibility) include prior experience in research, knowledge of the research field that interests you the most, the ability to find a supervisor, and an interest in graduate school or a research-related career. After reading through the eligibility requirements you will have to make a decision for yourself as to whether it is worthwhile that you apply to this program.

Providing advice on this topic is rather challenging as one student may be applying from a very different background than another. What I can share, however, is that if you do not necessarily meet all the eligibility requirements or qualifications for the summer studentship, you should not automatically assume that you are unable to apply. While highly competitive summer studentships have stringent and unforgiving eligibility requirements, others list strict requirements in the initial application so as to reduce the number of applicants they receive, though they may still consider students who have a healthy attitude but lesser experience.

A helpful strategy that I have personally employed when applying for summer studentships has always been to first inquire with the department about the strictness of their eligibility requirements when recruiting undergraduate research students. That being said, I never explicitly state that I am a student who does not meet the eligibility requirements, as this could leave a negative first impression on the coordinator responsible for distributing the research applications. I would advise that it is worthwhile to apply if you are close to meeting the mandatory eligibility requirements or if you lack one or two factors that are deemed an asset by the selection committee. It is not as uncommon as you think for students holding a B-plus average to gain acceptance to a program requiring a minimum average of A-minus. If your grades are any lower than one grade point below the required requisite, however, then I would suggest that a better use of your time would be to apply elsewhere.

With regard to not possessing certain qualities that are preferred in candidates, you should explain how you are still a competitive applicant in any written section of your application. In general, if you can defend why you are still a competitive applicant, than it is a worthwhile effort to apply.

The majority of summer studentships are not only prestigious, but they are also generally well-paid positions. Unfortunately, as a result, this makes them a highly competitive form of undergraduate research that places a particular emphasis on academic achievement.

Common Application Questions

After you complete your personal, contact and educational information, most applications also ask the student to answer a series of written questions. While some responses may be simple and only require you to re-word certain portions of your cover letter or curriculum vitae, others may present you with greater confusion.

For example, a common question may ask you whether there is a certain field in which you would prefer to work during your summer studentship. Always provide a clear and direct answer, while also stating that you would be willing to consider other fields if positions within your preferred field are unavailable. This shows that you are certain of your interests, but also willing to be flexible in the event that there are no field-specific positions available to you, and thus increases your chances of acceptance.

Another common question asks about your career goals. Applicants will answer this question in many different ways, however, I would advise that you answer this question by relating it to the research groups in which you plan on applying. It would not be wise, for example, to state that you are interested in becoming a health professional if you are not applying to a health-related research program, nor would it be wise to state the vice-versa. Always think about the educational backgrounds of your admission committee. For example, if you are applying to a summer studentship in cancer research, it may help you to state that you are interested in becoming a clinician-scientist, since many professors hold both MDs and PhDs and are clinician-scientists themselves.

Finally, one last question that is often found in research applications asks you to explain what interested you in the particular program that you are applying for. Again, you should try to relate your response to the field in which you are most interested in, and explain how your life experiences relate to this interest. It would also be beneficial if you are knowledgeable about some of the significant discoveries made within your field of interest and use these as examples of why you see potential in studying the subject.

In summary, one or more of these three questions are commonly found in undergraduate research applications, and they are often asked because supervisors know little about undergraduates, unlike graduate students who complete far more detailed applications. In addition, many supervisors are curious to see whether undergraduate applicants are certain of their research interests yet. Keeping these suggested strategies in mind, it is important for me to clarify that I am not advocating that you lie on your application. My aim is to provide strategies that can help the student who has a genuine interest in a research program to gain acceptance. As I mentioned earlier in this book, you should always be striving to create a series of research experiences that mirror your professional endeavours. As a result, if you feel that using my strategies would lead to dishonesty on your part, take this as a strong indication that you are not pursuing the right research field and that you need to reassess your academic interests.

The "Research Proposal" Question

If the department's summer studentship policy requires you to find a research supervisor first, sometimes the application form will ask you to briefly outline your proposed research project. This is where you will likely require your professor's assistance, since he or she will usually be the one who provides you with your summer research project. Writing a research proposal is not as simple as writing an essay for one of your undergraduate courses, and being able to complete a well-written summary takes much effort and practice. I will discuss this topic further in Chapter 11, where I will explain in detail how to apply for undergraduate research grants. While research proposals for the purpose of application are usually used to simply determine that you have a summer project, proposals for research grants are used to assess whether your project is worth funding, making this process extremely competitive. As a result of learning how to write an exemplary research grant proposal, the task of writing an application research proposal will become simplified. It is for this reason that I encourage you to perfect the skill of grant writing before you apply for your first summer studentship. Before you are able to write your research proposal, however, you will first need to know how to prepare for the interview with a potential supervisor, and this is exactly what will be covered in the next chapter.

CHAPTER 7
Preparing to Make Contact

Receiving Interview Invitations

Gaining an interview invitation shows that you have successfully demonstrated to a potential supervisor that you are both capable and interested in involving yourself in undergraduate research, however, do not congratulate yourself just yet for now you must prepare for this first meeting. The first impressions you make on a potential supervisor can either help you to secure that well-sought after position, or completely destroy your chance of gaining a research experience all together depending on what you do (or not do) before, during and even after your interview.

Pre-Interview Preparation

In this chapter, I will discuss in detail what you need to know about the interview process, how you can prepare for it, and what you can do to make a good impression on your professor during the interview itself. By this time, I assume that you have already diligently responded to your professor's reply thanking him or her for the opportunity for an interview and confirmed a date, time and location to meet. I assume that you have now also ensured to ask your professor whether you should bring anything to the interview, and have packed copies of your cover letter, curriculum vitae and academic transcripts in the event that you need them.

Reading and Understanding Research Publications

In Chapter 4, I explained what you can learn from a professor's authorship number on an academic publication. As you prepare for your interview, this information will become particularly important. The best way for you to understand a professor's research interests are by reading up on research papers published by his or her research group. You can accomplish this by doing a search of the professor's name on a scholarly database. During your search, you should pay particular attention to primary research articles, especially those that have been primarily funded and supervised by the professor and contain authors who are members of your professor's research group. Please review Chapter 4, if you would like to remind yourself about authorship numbers and how to identify projects primarily funded and supervised by a professor of interest.

Focus on reading the four or five most recent primary research articles that fit this description, as this will help you to anticipate what types of projects you may be involved in if your professor offers you a position. Reading these research papers will also help you to gain a better understanding of the protocols and technologies used in your professor's research group. Research articles are understandably often very complex, and cannot be read like this strategy manual, or even an academic textbook for your undergraduate courses. Read each sentence of a research article slowly and carefully, and try your best to identify key themes and repeated concepts in each paper.

When I prepared for an interview for an undergraduate research position, I highlighted all the scientific terms and phrases that I did not understand, compiled a list, and then looked up definitions online. You may also find it helpful to read about different technologies that are unfamiliar to you on an online science encyclopedia as well. This will help you to better comprehend how and why protocols were developed for published research experiments. I would also advise that you bring these research articles to your interview, so that you can share with your professor that you have already read about his or her work. Not only will this demonstrate enthusiasm and preparedness, but you can also use this to ask intelligent questions about your professor's research if the situation arises to do so.

Remember that if you choose to accept the research position, you have just reached the beginning stages, so do not treat this step as unimportant. If you are able to understand the content of research articles now, it will save you much time later when you are actually participating in conducting research experiments yourself. Furthermore, you can expect that you will need to read up on many more research articles both written by your professor and by his or her collaborators or colleagues, as the field of research is always growing, expanding and updating itself.

The practice that I gained reading research articles helped me in many different ways throughout my undergraduate career. By honing my ability to read and comprehend research articles efficiently, this allowed me to better plan my experiments, study for my courses, teach students and apply to graduate school all more effectively than had I not practised this skill throughout my undergraduate research experiences. Whether you are reading research articles to prepare for an undergraduate research interview or updating yourself on the latest discoveries in your professional career, this skill is invaluable because it is relevant regardless of your career goals.

Meeting Research Personnel

Another useful, but admittedly, more ambitious goal is to meet with a professor's research personnel or colleagues before your interview with the professor himself. When I applied to graduate school, I made it a goal to familiarize myself with as many researchers as possible within the specific field I wished to study. Whether it meant emailing, calling or meeting new connections in person, I took the opportunity to ask them about what they thought of professors whom I was considering as potential graduate supervisors. While this is more pertinent to graduate studies, it is not a bad idea for you to apply this method to your undergraduate research application process as well.

Professors, graduate students and other research staff in the same research field are your greatest source of information if you want to know more about the professor who will be interviewing you in the coming week. Do not be afraid to meet with these people and ask them what they think of your professor and his or her work. People who work closely with your professor, especially graduate students, are usually more than happy to share any information regarding their professor's work-ethic, personality, interests and perhaps even pet peeves, all of which you can use to your advantage before the interview even begins!

Articulating Your Thoughts and Ideas

Finally, the last step to preparing for your interview is highly personal, as your ability to articulate your thoughts and ideas clearly and concisely will largely determine the time you will need to adequately prepare. In other words, if you can answer unexpected interview questions both rapidly and intelligently, then you will not need as much time to prepare, compared to someone who typically feels a sense of nervousness in such a situation.

We all react to unexpected questions differently, and we all improve our ability to react with a clear and concise response at a different pace. Personally, I never felt that this was one of my strongest skills, and so as I continued to gain different undergraduate experiences that forced me to meet people and answer questions, I began to improve slowly and steadily. The best advice that I can give you is to practise with a friend who is also actively applying for a research position, because in some sense, it replicates the experience. The interview cannot be avoided, thus it is far more reasonable for you to diminish your fear of being interviewed instead.

In today's competitive world, the interview can either be your greatest enemy or source of salvation whether you apply for undergraduate research positions or teaching assistantships, further education such as graduate or professional school, or your dream career. Start off by carefully brainstorming a list of questions that your professor may ask you during your interview, but do not worry about how you will answer them just yet. You can also search online or at your university's career advising office for sample interview questions pertinent to undergraduate research positions. I have included a comprehensive list of questions that you can use to prepare for your interview, found in Section 1 of Appendix B.

Once you are satisfied with your list of questions, practise asking and answering questions with your friend. Pick the questions that you believe your friend may have difficulty answering and likewise ask your friend to do the same for you. After answering a question, ask your friend to provide you with feedback specific to your response, such as your content, tone, gestures, and ability to articulate. Do not worry about every little detail regarding your interview ability, as forming a comprehensive response is a skill that will take time to develop. You cannot and should not expect that your performance will be flawless if you are preparing for your very first interview. Instead, discuss what you did well, but more importantly discuss what you did not do so well, or what you could have included (or excluded) in your response. My philosophy has always been to not be afraid of allowing your friend to criticize you harshly because it is ultimately better that you make your mistakes in front of your friend than your professor, as constructive criticism is a small price to pay for not losing the research position to a more competitive candidate.

Finally, try to schedule all of your interviews within the same time period so that you can compare the offers that you are provided with. While preparing for multiple interviews simultaneously can be stress-inducing, it is the only way I know of that allows you to obtain the best possible offer without making your potential supervisors wait needlessly for you. If you are able to read up on your professors' latest academic publications, meet some colleagues and graduate students and practise a few interview questions with your friend before your interviews, I would say you are in good shape. In fact, you will likely be in better shape than the majority of your competitors and you should now be ready to attend your interview.

CHAPTER 8
First Impressions: The Interview

Research Interviews are Not All the Same: A Series of General Guidelines

I want to begin this chapter by first stating that interviews vary greatly from supervisor to supervisor, with regards to content, length and format. These interviews are not standardized like those of an admissions committee to a formal undergraduate degree program, and so you should be prepared to expect variance in style and format when being interviewed by different professors.

Making a Memorable First Impression

After you have confirmed an interview time with your professor, be sure to show up on time and dress appropriately. Your appearance is something that your potential supervisor will be able to assess even before shaking your hand, so make it count. While you may want to dress in your best clothes when meeting all your professors, it is also important that you consider the work environment. For example, if you are applying for a lab-based position, you should expect that you may be given a tour of the facility, hence you will need to adhere to a certain laboratory dress code. If you are instead joining a research group that involves patient contact, make sure that you are aware of the hospital dress code. Regardless, dress neatly and professionally to show that you are eager to begin your research position.

Begin your interview by thanking your professor for his or her time and be as friendly and polite as possible. Try not to be over-talkative, but instead realize that there are certain things your professor will want to know about you. Let them ask all the questions they wish to ask before asking too many of your own.

While you likely will not be asked any challenging scenario-based or behavioural questions, you can expect to answer questions about your personal interests and character. For example, it is common that your professor will ask about your interests in their research program, your work-ethic and other commitments, the types of experiences you are wishing to gain and your future academic or career pursuits. While it is impossible for you to prepare for each and every possible question your professor may ask, I cannot emphasize enough that it is important for you to be able to articulate your thoughts and ideas both clearly and concisely.

The interview is also the time when your professor may ask you for any or all of your application documents, however, it is important that you do not mention anything about or produce these materials unless you are asked specifically. Many professors will not assess your academic history, but instead, trust that you are a hard-working and academically successful student. If you are well-prepared, this portion of the interview should not be too difficult for you, as after all, you are sharing your interests and future endeavours with someone who may soon become your mentor. Some professors may cover this section briefly, and instead focus on spending the majority of the interview explaining potential research projects that you could work on as a research student.

Undergraduate Research Projects

While few professors will ask questions directly testing your knowledge of their research, they may expect you to have a basic background of their work so that they know that you can make an informed decision when choosing your project. Undergraduate research projects come in many different types, and it will be up to you to determine which type of project suits you best.

If you are serving as a part-time volunteer, your ability to commit yourself to a research project is likely to be very limited. In this case, you may be an assistant to your professor or a graduate student in the research group. This can be useful, especially if you do not have any background in research, as you will learn how to complete simple tasks such as creating a stock solution, running software programs or editing a survey package. It is in this way that you will learn how to prepare for experiments. Your duties may also include helping the research group with minor housekeeping tasks such as updating a records book, cleaning equipment, or taking an inventory.

In contrast, if you enroll in a senior thesis or make a long-term commitment to a research group, you may be fortunate enough to acquire your own research project. This means that you will be responsible for the designing and planning of your experiment, and you may even need to write out a grant proposal or submit your study for ethics approval. In this role, you will learn all the necessary protocols associated with your study, and if you are fortunate enough, your professor may even allow you to supervise one or two less experienced undergraduates in the future.

For the most part, however, most students conducting research at the undergraduate level will be involved in working on a study that is a smaller portion of a much larger project led by a graduate student or even the principal investigator. While you will still be responsible for understanding the experimental protocols, your experimentation may be guided by a more senior member of the research group.

Regardless of the type of project you are offered, listen closely to what your professor shares with you, understand the nature of the research study being offered and ask questions if you do not understand something. This is also the ideal time to ask any questions about your professor's research publications that you had read before the interview. While this will show your professor that you have made your best efforts to prepare for your interview, do not spend too much time asking irrelevant research-related questions as you may then be perceived as too self-reliant on others for help.

Be Wary About Accepting an Offer on the Spot

If the both of you can collectively decide on a research project and your professor is pleased with your interview responses thus far, it is often the case that you will be offered a position on the spot. If this is not the case and your professor explains that they are interviewing a few scores of students, be wary of accepting such a position even if you do receive an offer in the coming days. These professors usually recruit a few undergraduates during one term or summer and so you should be aware that while this process may be competitive, you also may not have as many opportunities to develop a solid professional relationship with your supervisor. Always ask your professor how many undergraduate students he or she plans to hire in advance of accepting a position.

Even if you are offered a position, I would advise that you wait a short time before accepting, unless you are absolutely certain that this is (and will be) your best offer. If you are unsure, do not rush the acceptance of an offer especially if your professor does not need to know your answer right away. As I mentioned before, it is important that you are certain that this position suits your interests and needs, but it is also equally important that you accept a position only when you know that it is (and will be) your best offer.

Asking the Important Questions

The final portion of your interview will begin when your professor has exhausted their list of questions for you, and will then ask you if you have any questions for him or her. Often many undergraduates do not recognize that this is a valuable component of their interview and make the mistake of ending the interview without asking any questions. From a student's perspective, this is actually the most important component of the interview as you can now freely ask questions that can help you to determine if this is a position you wish to accept. By stating that you do not have any further questions, this may indicate to your professor that you lack interest in their research program, or that you are not knowledgeable enough about the position you applied for, both of which may result in your professor withholding an offer to you.

I am always most prepared for this question, as this is the portion of the interview where I can freely investigate whether my needs and interests would be met if I accepted the position. Before I attend any interview for a research position, I always prepare a series of memorized questions that are essential to my investigation. If you do not wish to memorize anything, make a list, but it is essential that you are prepared to ask your potential supervisor some necessary questions. While some of these questions may seem trivial, it is still important that you are on the same page as your professor in order to avoid unexpected and unpleasant surprises in the future. You can find a list of all the important questions that you should ask every potential supervisor in Section 2 of Appendix B.

In the Event of Rejection

Before I share strategies with you regarding how to decide which offer to accept, I feel that it is important that I address what happens if you are not successful in gaining any offers following the interview process.

It is Not Always Your Fault

This can be difficult for students, especially if they have already put in a lot of work towards securing a research position, however, it is important that you are aware of the reason or reasons why you were unsuccessful. Realize that sometimes it is not your fault, as some professors received applications from students who possessed more research experience and education than you.

Re-Assess and Re-Apply

After you receive notice from your professor that you were not offered a position, think about all the possible mistakes you could have made so as to improve your performance in the future. Perhaps you were unable to answer a few questions clearly or confidently, or maybe you gave your professor a poor first impression. Because you were offered an interview, it is highly unlikely that your professor found any outstanding issues with your original application package. Instead, you probably made your mistakes during the interview.

I would suggest that you email your professor expressing your thanks for their time, but also ask why you did not receive an offer and how you could improve yourself as a candidate in the future. You may also want to consider asking your professor whether he or she knows of any other colleagues hiring undergraduate research students if you had not already mentioned this in your original application email.

The positive side of this is that you can continue to apply for other research positions, and if you had prepared a secondary list of professors to apply to, all you need to do is send out a few more emails. While applying during certain periods of the year is more likely to land you a research position, do not forget that most professors accept applications and hire students throughout the year.

Accepting an Offer

If you have received multiple offers after attending all of your interviews, it is important that you quickly decide on a supervisor. If you have no prior research experience and only received one or two offers, the answer may be very clear, however, if you have received more than a couple offers, the decision may be more difficult.

After attending all your interviews, you should take the time to carefully reflect on the positive and negative aspects associated with each offer you received. The truth is that your undergraduate career is short, and the four or five years will go by quicker than you think. You cannot afford to waste your time and resources on a research experience that provides you with little opportunity for learning and advancement.

Not all Offers are Good Offers

When making your decision, do not accept offers from professors who do not support your future endeavours during the interview or who expect you to plan too far ahead in your academic career. While the reasons for this are obvious, year after year I have seen students accept offers out of desperation or because they feel that the quality of their professor is less important than his or her research program.

Unfortunately, doing this can lead to further difficulties, as the quality of mentorship you receive from your professor will ultimately determine whether or not you will be able to form a long-lasting professional relationship with him or her. Additionally, if you feel that you gained a research experience of poor quality, you may subconsciously lose interest in the field that initially fascinated you, even though it may still hold potential for your future education or career.

During your interview, be cautious of professors who ask about your future endeavours and are dissatisfied by your response, provided that you answered honestly. For example, if you explain that your end goal is to gain acceptance to a professional program and your potential supervisor tells you to pursue graduate studies instead, or if your potential supervisor states that you must make a commitment to a graduate program if you accept an undergraduate research position with him or her, my best advice would be to decline the offer, unless you absolutely have no other options.

Sometimes professors will openly share that they are unwilling to provide research positions to students who have certain goals, and yet I still know of students with those exact goals who have lied to their professor in order to gain the research position.

I can safely conclude that this nearly always ends in disaster, for numerous reasons. If your professor finds out, he or she will lose all respect for you, but if you are somehow able to conceal the truth, you will still face a series of negative consequences. The most obvious is the fact that your supervisor can never serve as a reference for your academic goal since you would need to admit that you had originally lied, all while you knew that he or she does not support your endeavour. Furthermore, if you end up disliking the specific field of research, it is more likely than not that your professor will not support your transition into exploring another field since he or she was under the impression that you were willing to commit to a graduate school or career within the same field.

Despite what I have just mentioned, understand that there is nothing wrong with your professor questioning the reasons for your endeavours, or explaining the obstacles, limitations and controversies associated with your desired future field or program of study. In fact, this may actually be helpful as you will be forced to defend your interests, explain why they are important, and think about whether you are pursuing your goals for the right reasons. An issue only arises if your potential supervisor is unwilling to support your goals all together, or asks that you make an academic commitment that is too large or long-term. If this is the case, then I would highly advise that you try to seek out an alternative offer that better suits your needs.

The Supervisor Screening Method

Perhaps this is overly critical, but I would suggest that you assess the behaviour of your potential supervisor knowing fully well that your professor is directing 100% of his or her attention towards you during your interview. This may not be a necessary step if you only receive and attend one or two interviews, however, once you advance in research and gain multiple offers, I am certain that you will find the following strategies useful.

Firstly, I strongly advise against accepting positions with professors who are late to your interview. Through my own research experiences, I have learned that a professor is not fully committed to the task of supervising you if they are very late or absent to your interview without providing you with advanced notice, despite the fact that they extended the invitation. It does not matter what their reason is for being late or absent either, for whether they were in a meeting, running an experiment or doing something else, they did not consider your interview to be of as much importance, and I can almost guarantee that they will not consider your future meetings with them to be of much importance either. In contrast, if your professor informs you in advance that they will need to reschedule the interview, take this as a sign that they experienced a true emergency, yet ensured to remember you in the midst of everything.

Personal experience has taught me that professors are very habitual people, and they need to be because of the nature of their job. The nature of the career itself is challenging, stressful and demanding, and each professor will react a little differently. After meeting with and working alongside multiple professors, you will begin to see that those who arrived late to your interview will also be those who will show up late to meet with you in the future, while those who show up prepared will also be those who will ensure that all your deadlines are assigned and met.

Since you likely have not experienced this first-hand, I expect that you are now wondering whether there is a way of predicting the quality of your potential supervisor's mentorship before accepting the position, and the answer is that you can. The procedure is actually quite simple, as supervisors are usually very straightforward when sharing their commitments. During the interview, ask your professor direct questions regarding how much time they will have for you throughout the duration of your research term including, whether they will be travelling abroad, whether they have other commitments besides research (i.e. teaching a course, serving on a governmental or academic committee, etc.), and finally, who you can consult for assistance if they are not around. By knowing these answers before even beginning your research, you should have a good indication as to whether you will receive adequate mentorship during your research experience. It is also important for you to ask whether you could meet with them regularly to discuss any issues that you may have, despite the fact that you will likely need to learn a few things on your own as you will not have the opportunity to consult with them on a daily basis.

Understanding the Research Setting

While I am providing you with a screening method that you can use to determine whether your professor will serve as an ideal mentor to you, I also want to remind you that all professors will experience periods during which they will be extremely busy. Through providing you these strategies, I am in no way advising or condoning that you become overly reliant on your supervisor throughout your research experience.

In every research setting there will be times where you will have to consult academic papers, equipment instruction manuals, and graduate students or research assistants, and you should not expect your professor to answer each and every question that arises as you conduct research. Making a good impression on your professor definitely demands that you show initiative, but it also means that you can demonstrate to him or her that you can think for yourself while overcoming obstacles and solving problems you encounter in the research setting on your own. It is more important that your professor is satisfied with your performance, than that you are satisfied with his or her mentorship, because the former is what ultimately determines a long-lasting professional relationship and a valuable referee for your future.

If you have found the ideal supervisor and accepted a research position, I wish to congratulate you not only for gaining research experience, but also for all your hard work and preparation. I can assure you that this will increase your competitive edge in academia and beyond, but more importantly, this experience will reshape the way you think and make decisions in your life for the better. The vast majority of my work is now done, however, I have included four more chapters to help you continue on your journey after you have gained an experience in research. In Chapter 9, I will discuss the process of beginning your research project and how you can make the most out of your research experience.

CHAPTER 9
Optimizing Your Research Experience

The Beginning

It is perfectly normal to feel slightly overwhelmed on the first day of your research position. The facility, equipment, people and procedures will all be new to you, as well as the scientific jargon that you will hear being used around you. You can almost think of your research position as an enrollment in a new class, in the sense that a research group is also an experience in learning, except this time you will be working with many people who have two degrees or more. You may think that this is a little intimidating at first, but I am certain that you will eventually find that research provides you with the opportunity to practically apply the knowledge that you have learned in your undergraduate classes.

Within the first two weeks, you should be prepared to do far more reading than experimenting, as you will need to gain a background in research regarding safety training, equipment or program operation, and research protocols. During this time, it is also important that you meet with your supervisor to confirm your project and learn how you will accomplish your assigned tasks. Most professors will give you a series of academic publications for you to read in order for you to gain the necessary background needed to complete your research project. Often, the majority of these publications are review articles that comprehensively summarize the developments and discoveries that have been made in your specific area of study over the past few years. If your professor does not provide you with these articles, do not be afraid to ask him or her for these resources, as it is important that you understand the background behind the research you are conducting. Furthermore, you should also begin to familiarize yourself with acquiring these articles yourself by searching through an academic database, as you may need to retrieve more articles later if you are asked to prepare a presentation for a laboratory meeting or write a poster, abstract or paper at the end of your research term.

Working Efficiently and Effectively

Realize that the average undergraduate research term rarely exceeds three to four months, with the exception of an eight month senior thesis. Regardless, this means that you need to make the most of your time and work both effectively and efficiently if you plan on completing your project. As soon as you have familiarized yourself with your project's background, you should promptly begin planning for and conducting your research study.

Ask When You Need Help

Undergraduate research projects vary greatly from student to student, sometimes even under the supervision of the same professor, hence it is very difficult to provide you with advice specific to your field of interest. Keeping this in mind, you should take full advantage of the resources you have available in your research setting, whether it be a more experienced undergraduate or graduate student, your professor, or other research associates. If you do not have the knowledge or understanding of how to do something, it is always a good idea to ask first instead of completing a task incorrectly, or worse, unsafely. Unfortunately, in research, there are often no second chances when it comes to experimentation. In the natural sciences, the use of reagents or equipment is often very expensive, and so you may only have one opportunity to complete the protocol correctly. Similarly, in the clinical sciences, your research often relies on data from rare specimen samples or patient compliance, both of which are irreplaceable if lost or processed incorrectly.

Work Carefully

It is important that you work efficiently, yet carefully, and that you record all your procedures and experimental results in your department-assigned notebook. This is the only way to ensure that your findings are recorded permanently and can be reviewed by your supervisor even after you leave your research group. Furthermore, you will need to keep a neat and well-organized notebook of your findings in order to complete your report and present your project at the end of your research term.

Interact With Your Supervisor

It is also important for you to meet with your supervisor on a regular basis. Schedule meetings with your professor at the beginning of the term, and select a day and time that works for the both of you, so that you can follow this routine throughout your research term. If your professor is busy during a certain time-period, be sure to remind him or her of the meeting, and be willing to accommodate a request to re-schedule.

It is important that you regularly discuss your research with your supervisor for two reasons. The first is that it will help you to stay on track with your project, and ensure that you are receiving feedback in a timely matter so that you can make minor adjustments throughout your research period, instead of major adjustments right before your research experience ends. The second reason is that this will ensure that you can nurture a long-lasting, professional relationship with your professor, so as to gain a valuable reference, colleague or even collaborator, in the future.

If you never interact with your supervisor, he or she will not get to know you and may not even remember you one year after your research experience ends. You must always be confident to take the first step in approaching your professor because you are the less busy person, as well as the one who needs the benefit of a reference and an academic mentor. Unfortunately, many of my peers were unable to reap these two benefits due to a lack of interaction with their supervisor, so it is important that you do not make the same mistake.

Every now and then, you may also find it helpful to ask your supervisor for some feedback regarding your performance and professionalism. By this, I mean that you ask your professor how he or she feels about your character, personality and work-ethic. This can help you to determine where you should focus your efforts with regards to self-improvement for the future. This also allows you to make a reasonably effective prediction as to what your professor may say (or not say) about you if he or she serves as your reference.

Beyond assessing your strengths and weaknesses, lastly and most importantly, be punctual, dedicated and hard-working in everything you do in research. These are the three qualities that are easily noticeable for their presence and their absence. If present, these qualities will ultimately help you to successfully complete your project and leave a long-lasting and positive impression on your professor, not to mention the entire research group.

Additional Opportunities

If you want to make the most of your research experience, I would highly encourage that you take the extra step of seeking out opportunities that extend beyond your research project. Within your research group or department, take the opportunity to attend laboratory meetings, clinical rounds, colloquia, seminars or any other professional gathering that allows you to learn more about the projects that your fellow research group members are working on and better understand your field of research. Attending these events also provides you with the opportunity to network with people in the same field who may one day become your future supervisors or colleagues.

Of all these events, the laboratory meetings (or research group meetings, if you are not working in a laboratory) are of particular importance to you as an undergraduate. Sometimes professors do not make it mandatory for undergraduates to attend, however, there are numerous benefits that you can gain by becoming involved. For example, you will gain the opportunity to share your own research project and receive constructive criticism and feedback from your professor (and others) regarding your research project, presentation skills, and written work. Laboratory meetings also allow everyone in the research group to share their weekly progress, discuss academic papers that have been recently published by other researchers internationally known to the field, and share knowledge of any upcoming events relevant to the field of research.

Beyond this, there are often opportunities for you to attend research conferences or symposiums that showcase a broader spectrum of research, usually within one university department or multiple similar departments. Ask your professor if there are any events scheduled in the near future, and if so, find out how you can attend. Besides also being able to network at conferences and symposiums, you will also have the opportunity to view the work of research staff and other students. By viewing poster sessions or listening to oral talks at these events, you can take note of the best presenters and use them as a model for your own research project reports.

Overcoming Obstacles in Research

Regardless of how much you may prepare, it is also important for you to understand that even with the best research group and project, some hours, days or even weeks in research can be tedious, dissatisfying or just downright frustrating. I have experienced many of these periods during my time as an undergraduate, and these situations are not made better by the fact that the university lifestyle overwhelms students with university coursework and other academic commitments.

Even if you put the most time and care into your research project, you must still expect the unexpected, and often many setbacks become avoidable only through greater experience. On one of my busiest and most tiring days I had arranged to meet with one of my research professors who asked me to draw the cell cycle in front of him upon stepping into his office. A simple task, and something I could have easily drawn since my high school days, however, I crumbled under the pressure of completing this trivial task.

No doubt, some professors have been upset with me for reasons I may never understand, while others have questioned my career path and told me that I would find a lack of success there. Even if you set out to be the best student you can be, these things will happen to you too.

As a result, it is important that at the end of each (long) day you remember that none of this matters. What matters is that you can pick up from where you left off yesterday, rekindle your professional relationship with your professor and continue working on your project. If you can remain punctual, dedicated and hard-working, and exercise these three qualities consistently regardless of your obstacles that you encounter, you will greatly increase the chances that you will find favour with your research group and professor and successfully complete your research project.

Considering the Option of Resignation

Unfortunately, every now and then I hear of students who end up leaving an undergraduate research position prematurely because of a poor relationship with their supervisor due to varying circumstances. If you ever begin to think about resigning from your research position, I highly suggest that you make an informed and careful decision regardless of what you ultimately choose to do.

Deciding to resign is a personal decision that needs to be assessed on a case-by-case basis. On one occasion I considered leaving one of my positions, but ended up staying, while on another occasion I made the decision to resign from my position and consequently joined another research group instead. Regardless of your decision, there will be long-lasting and far-reaching consequences that you will need to consider, hence the need for discussion about this often undesirable topic.

If you leave your research group, you will lose any academic mentorship that you are currently receiving from your professor, as well the potential for a valuable reference. You cannot expect that your supervisor will be willing to speak highly of you in the future, nor would it be wise for you to trust them in writing you a confidential reference letter. Also, realize that your professor may share the fact that you decided to resign with his or her colleagues within the same department or research field. You will also lose any financial benefits that you are receiving from your position if it is paid job. Additionally, if your research position is a university course, you may face various academic penalties. Lastly, you may not have the opportunity to fully learn certain research skill sets, and any work towards your research project may be abandoned or forgotten, especially if there are other ongoing projects that are of higher priority to the research group. As a result, excluding leaving your position for medical reasons, I would recommend that you take all these factors into careful account before making your decision.

Yet, there are times that warrant that you write your letter of resignation. If you are absolutely disappointed with your research position because you are being abused in any way by your professor or research group staff, making your work environment very difficult, this may be a sign that you should be leaving. If you feel that you are receiving a severe lack of mentorship and your professor is not fulfilling the majority of his or her agreements to you, this may also be another legitimate reason to terminate your position. In other words, if you are absolutely certain that you have put in your best efforts and feel that you are still unable to form a healthy professional relationship with your supervisor and/or research group members, I would advise that you consider resigning from your position, despite the negative consequences that will await you.

Upon choosing to resign, many students face a daunting challenge, as they are unsure of how they should confront their supervisor. It is important for you to always remain calm and respectful, even if you feel that all faults fall on your professor or research group. Explain your feelings about the situation honestly and openly, but do not directly blame your professor for anything. Express your wish to part ways with the research group, but also be certain to thank your supervisor for their time, for providing you with the opportunity to conduct research with him or her, and for serving as your academic mentor. Despite the fact that this is an unfortunate situation for both you and your supervisor, it is always wise for you to also express that you wish to keep in contact moving forward, and that you are grateful for the opportunities that you have been provided. Unfortunately, writing a letter of resignation can be extremely difficult and can take a considerable amount of time, even if you do not have any left to spare! In the event that you ever find yourself in this situation, I have provided you with a resignation letter template in Section 5 of Appendix A.

However, with this information in mind, you should remember that in the vast majority of cases, if you have properly discussed the details of your research position with your professor during your interview, you should expect that you will enjoy your research experience and remain a member until your position ends as originally stipulated.

One extremely important point which I wish to make, however, is to never leave a research position simply because your assigned research project is not providing you with successful results. Resigning for this reason is not worth the negative consequences, as there is never any guarantee that another research project in a different research group will lead you to greater success. Provided your relationship with the members of your research group are healthy and you enjoy the work environment, it is important that you avoid any temptation to leave your current research group for another, as you will see in the next section.

The Outcome of Your Project Does Not Matter

The honest truth about research at the undergraduate level is that it does not matter how your project turns out. Many students are shocked to learn this fact because they have been taught to believe that their research project will turn out similarly to their high school science laboratory experiments, yet the difference between these two things could not be any greater.

As mentioned before, high school experiments are tried, tested and true, which means that they are designed to result in a specific outcome that is already well-known and highly understood by scientists. This means that students only conduct these experiments to learn about the research process, and not to make a novel discovery.

Conducting research in a real-world laboratory or clinical-setting means that you are investigating what is not understood, attempting to find the answer to unanswered questions, and designing and conducting experiments that no one else in the world has ever performed before. This is what makes research exciting, yet it is very common that a research project will fail to produce the results that you, or even your supervisor, initially anticipated.

There is no guarantee that by switching research groups you will gain better results, and it is highly likely that you will run into similar obstacles with a new research project too. Instead, focus on learning how to be a good scientist by perfecting your research methods and listening to the advice provided by your more experienced mentors. Any research project, if conducted properly, will always result in some finding. While it may not be the finding you anticipate, there will always be something that you will be able to write a report on, and at the undergraduate level it is more important that you learn how to compile and present your findings properly, than it is for you to make an earth-shattering discovery.

Honesty is the Only Policy

It is important that you conduct your research honestly, and there are no exceptions to this rule. While it is very common for students to misrepresent their findings and skew data in high school or even undergraduate courses, this is absolutely prohibited in the real world of research. It is important that I make you aware that researchers have been stripped of their professorship, fined and even sentenced to jail in the past for misrepresenting their findings.

While it is highly unlikely that you would face such harsh consequences, it is necessary that you understand that the information you include in your research notebook is a legal document, which can be used to verify your research group's recordkeeping in the future. Besides the risk of your supervisor terminating your research position, understand that misrepresenting your findings, if undetected, has direct repercussions on the way knowledge is shared, translated and applied in the real world.

Presenting Your Work

Upon completing your research project, you will likely have the added benefit of presenting your research work towards the end of your term. This is a great opportunity for students to experience what it is like to present a poster or provide a talk on a real-world application of science, and hone their communication skills for future use. Many students are very nervous about presenting their work, and are even more nervous about answering any questions researchers in the field may ask them. If this is the case for you, all it means is that you need to gain more practice through experience.

For your benefit, I have provided you with a list of helpful tips regarding improving your communication skills and constructing an effective research poster and talk in Section 3 of Appendix B. Read these over carefully as part of the preparation process, but also be sure to consult your research group, or even ask your professor if he or she has time for a practice presentation. You may also find it helpful to present in front of your friends who know little or nothing about your research field and project. Have them ask you questions at the end of your presentation, as this will prevent you from overlooking the very fundamental and trivial questions that you may not consider otherwise. I often practised presenting my research with the professors I worked for, as well as with my friends, as both helped me to prepare tremendously.

The Research Project: A Final Note

You should now have all the general information you need to be successful at conducting research and completing your own project. Needless to say, your research experience will eventually come to its end, sometimes too quickly, especially if you work hard and thoroughly enjoy yourself during the process. What some students do not realize is that leaving a research group poses its own unique challenges, just like any other step of the undergraduate research process, and it is important that you find successful ways to maintain contact with your research group, and in particular, your supervisor. The information in the next chapter is dedicated to exploring this issue.

CHAPTER 10
Maintaining Professional Relationships

Plan Ahead

While choosing the length of your research term is usually a very simple task, planning how to maintain positive and meaningful contact with your supervisor after your position ends often presents numerous challenges for undergraduates. Unfortunately, many students leave their position, add a new entry to their curriculum vitae, and only re-contact their professor two or three years later when they are in need of that elusive reference letter. This is by far one of the poorest decisions you could make with regards to maintaining a professional relationship with your supervisor, as I will later explain in this chapter.

Before your position ends, it is important that you plan to remain in contact with your professor, regardless of your future plans. Before you set up your last meeting with your supervisor, you should take the time to reflect over your research experience. Perhaps you may decide to ask your supervisor to extend your research term, either by continuing an old project or starting a new one. Instead, you may feel that you would like to begin another research experience in a different field, or perhaps take a break from the research world all together. Regardless of what you plan to do, make sure that you have come to a conclusive decision before discussing your future plans in research with your supervisor.

To Stay or Not to Stay

I advise that you set up your meeting so that it is both in person and in private, as opposed to during a laboratory meeting as you may not wish to share certain things with the other research group members. If you are interested in staying as a research student, you should ask your supervisor about his or her plans for the next coming months. While some supervisors are more than willing to let you stay, others may find this unfeasible due to financial or circumstantial reasons. For example, if your supervisor is planning to hire a new graduate student or lacks adequate funding to support his or her work, your ability to stay may be hindered significantly. Also, consider that if you have spent a summer studentship at another university, you likely will not be able to stay while concurrently completing your undergraduate classes at your home university.

Both staying and leaving have their own unique advantages as well as disadvantages, and so if your professor does happen to allow you to stay, consider the best decision for you. Staying with the same research group is advantageous because it allows you to work on a project for a longer period of time. Not only will you be able to better hone your research-specific skills, you will also have the opportunity to provide greater contributions to any given research project and increase your likelihood of gaining authorship on an academic research publication as a result.

If you renew your position into a busier time period during your undergraduate career, however, it is important to consider your ability to commit your time to research. For example, if you began your position at a time when you did not have any other commitments, and then extended your research term into an academic term filled with university courses, you risk jeopardizing the relationship you had formed with your research group and professor if you become too busy and neglect your research work. That being said, if you began your research term during the academic year and wish to extend your stay into the summer, you may have the opportunity to experience the vice-versa.

In contrast to staying, while moving to another research group requires you to leave your old project and professor behind, a new experience also provides you with a new supervisor and one additional future letter of reference, provided that you maintain a healthy professional relationship with both your past and current supervisors. A new position also allows you to expand your research skill sets, and explore different academic fields during your undergraduate degree.

One final option is to renew your current research experience as a volunteer position, thus requiring you to make a lesser commitment while allowing you to concurrently seek out and work with a second research group.

Regardless of your decision, you should tailor each subsequent research experience to more closely match the field that you are interested in pursuing, such as a further undergraduate research experience, or even graduate or professional school. Take advantage of the fact that you carry prior research experience, since this will provide you with the benefit of choosing from more opportunities. This is because at the undergraduate level, the sole fact that you possess research experience will allow you to gain a greater number of offers, regardless of the field in which you are studying. As an undergraduate, professors understand that you will need to explore your interests, however, they prefer to recruit students who have a history of commitment to tasks as lengthy and arduous as a research project. I personally employed the strategy of rarely renewing a research position, but ensured that I worked very hard during each experience, which ultimately provided me with the opportunity to learn many skills, meet more professors, and practise presenting my work more frequently. Eventually, you will have to find your own unique strategy that supports your personal interests, personality and work-ethic.

Preparing for the End of Your Research Position

Regardless of how many times you renew any position, it is expected that you will likely leave at least one research group before your graduating year. As a result, you should ensure that you also receive information about how you can contact your supervisor in the future by planning to meet with him or her shortly before you anticipate your research position to end.

Be sure to thank your professor for providing you with the research opportunity and express that you would like to maintain contact with him or her in the future. This ensures that you are sending a clear message that shows you value them as a mentor and that you are making the effort to keep in touch periodically instead of only showing up when you need their assistance.

A strategy that I highly recommend is for you to ask them that they write you a general reference letter that comments on your academic ability, personal character, and aptitude for research directly following your departure from the research group. You should explain that it will be used when you apply for your next research experience, non-research job or extracurricular activity. Besides retaining a general reference letter for future applications, it also serves as official documentation that you conducted research under the supervision of your professor, without having to re-verify with him or her if you are ever asked for documented proof. Finally, and most importantly, obtaining a general reference letter allows you to obtain your professor's professional opinion of you.

It is for this very reason that some professors refuse to provide you with a reference letter until you need it, so that their thoughts are truly communicated confidentially. This can make it more difficult for you to know what your professor may write about you, however, an alternative option in cases such as these is to ask them about what they would write in your reference letter. Regardless of whether you are able to obtain a general reference letter or only talk about one with your professor, use either opportunity to assess your strengths and weaknesses. Know your strong points, but more importantly assess your shortcomings and focus on understanding how and why you should correct them. This is the only way for you to effectively improve your performance in the future.

Additionally, you should ask your professor whether his or her research group will be holding any events that you can attend in the future. You should also strive to visit your professor at least once per academic term if you wish to maintain a continued professional relationship. You do not necessarily need to visit for any particular reason, but instead take the opportunity to share how your university experience has progressed and find out about the progress of your professor's research group. You may also want to consider providing your professor with a thank you note or even the occasional small gift for helping you gain experience in research and for serving as your reference.

Asking for a Future Letter of Reference

Eventually, there will be a time when you will need a confidential reference letter from your professor, and it is important that you make this request appropriately.

Regardless of whether this letter of recommendation is for graduate school, professional school, or something else entirely, be sure to contact your professor at least a month (two if possible) before the application deadline. The greatest disfavour you can do for yourself is to ask for a reference letter at the last moment, because your professor may write it in a rushed and careless manner. Making a last moment letter request also puts unnecessary stress on your supervisor due to their many other commitments, and it may even be the case that they tell you that they will be unable to complete it for you in time.

If your supervisor agrees to serve as your referee, be sure to send them all the details regarding what your reference letter needs to contain, such as the required content, page length, deadline, type of application (i.e. Master of Science, Department of Biology), and all the instructions regarding how and where to submit your letter. Specify any other responsibilities required of your supervisor, such as completing a separate confidential assessment form, acting as a verifier, answering an email/a phone call, etc.

Do not be afraid to forward your professor your updated curriculum vitae and make a request as to what you wish they would comment on in your letter, whether it is a certain characteristic, skill or academic or non-academic achievement. While some professors may prefer to write their letter without taking your requests into account, it is always best to make a request anyway as there is nothing you can lose by doing this.

Finally, be sure to thank your professor for serving as your academic referee, and keep them updated on the status of your application.

CHAPTER 11
Advancement in Undergraduate Research

Use Subsequent Research Experiences to Reflect Your Professional Goals

While an exceptional academic record and ability to plan may allow a few students to gain their first research experience in a field relating to their future goals, it is more often the case that students will hold an unrelated experience simply to gain an undergraduate research background.

This chapter focusses on everything you need to know to advance in your undergraduate research career. In both Chapters 4 and 10, I explained the importance of tailoring your future research experiences to reflect the goals you wish to achieve following your undergraduate degree. Considering this becomes particularly important for students who have recently finished (or are about to finish) their very first research experience, provided they have additional undergraduate years to complete.

Usually most students will have at least broad idea of where they see themselves in the years that follow their Bachelor's degree, however, it is not uncommon for undergraduates to express uncertainty about their future. This uncertainty can complicate the planning process with regards to securing future research experience, hence the reason why you should keep your interests broad. For example, if you wish to pursue a career in healthcare, but are unsure of an exact profession, you could focus on gaining your second research experience within the field of health care, perhaps in a wet-lab environment, assisting with a clinical trial, or even working in the field of medical sociology. Instead, if you are more interested in environmental protection, you could instead focus your efforts on spending your second research position conducting fieldwork that promotes animal conservation.

If you follow this model you will find that as you progress through your undergraduate career, your future goals and your ability to gain research positions will both become more specific. For example, in your first year you may have an interest in health and healthcare, in your second year you may gain a specialized interest in therapies in cell biology, in your third year you may narrow down your interests even further to focus on molecular therapies used to treat cancer, and finally, in your fourth year you may decide that you wish to pursue graduate studies in molecular therapies specific to brain cancer.

As you can see, it is much easier if you align your undergraduate research career with your professional goals in your first year as opposed to your fourth year. There are many more opportunities available to gain research experience if you have broad interests, such as in health and healthcare, as opposed to specific interests, such as molecular therapies in brain cancer. Additionally, if you start your research career in your first year, by the time that you are in your fourth year you will have gained more focused training in specific and relevant research skills within your unique field of interest. In other words, this is a useful strategy to employ because it allows your research career to advance at the same rate as your undergraduate career.

In fact, if you are a well-planned and ambitious student, you should make efforts to advance your research career at a faster rate than your undergraduate career, provided that you have decided on your post-undergraduate career goals. For example, if you know in your first year that you wish to pursue graduate studies in molecular therapies in brain cancer, then you can focus your efforts in pursuing research experiences specific to only this field, as this will result in a curriculum vitae that shows any admissions committee that you are a student who has been historically certain of your goals.

However, it is also important that I share a word of caution, because if you change your professional goals during your undergraduate career, this may result in you having to drastically change your research interests as well. This can be disadvantageous as your past research experiences may reflect your indecisive past. Of course, you can always remove unrelated items from your curriculum vitae and choose to never mention them to any potential employers or admission committees, however, this still results in a disadvantage because you cannot showcase as many activities as you have actually completed. As a result, it is extremely important that you only begin to specialize your research interests when you are absolutely certain of your future career and/or path in education.

Undergraduate Research Grants

Understanding the Value of an Award

If you so choose to advance in undergraduate research, you may also wish to consider applying for grants. You may be wondering what purpose grants serve for benefitting undergraduates, as after all they do not fund a project in the same way as a professor's research grant. As promised in Chapters 2, 3 and 6, I will use the next few pages of this chapter to discuss undergraduate research grants in greater detail.

The application process for undergraduate research grants are highly competitive, which makes gaining such an award both impressive and prestigious. These grants are awarded through either public governmental organizations or private research foundations. While most researchers at the principal investigator level place greater value on grant funding from public governmental organizations, at the undergraduate level you typically do not need to worry about this issue unless you are aware that your funding body is promoting research conducted in a biased-manner. Any respected research grant, even at the undergraduate level, should strive to select their successful applicants using a fair and impartial process, therefore, it is important for you to read about how each granting body receives its funding to support students, as well as how the application and selection process works, before you consider applying.

The most prestigious undergraduate research grants are provided by governmental organizations, councils or institutes, followed by private non-profit research foundations that primarily rely on the support of their donors. Be more cautious if you choose to apply to for-profit private research foundations or granting bodies affiliated with or supported by a company, as funding from both these sources may support biased research due to the granting body's for-profit nature. Because many for-profit sources of funding often require the researcher to promote a certain viewpoint or finding, this can result in a declaration of a conflict of interest during the publication process. It is for this reason alone that many researchers working for an academic institute, such as university professors, strictly refuse to be funded by for-profit granting bodies.

Research Grant Application Strategies

I am able to provide you with a unique perspective as I have received one undergraduate research grant from a governmental council and another one from a private non-profit research foundation. The application process for grants may vary from one to another, however, many commonalities exist between the application processes, and if you wish to be a successful applicant you will need to develop certain skill sets and possess some standard academic characteristics.

Primarily, you will need to score exemplary grades, in fact, typically the top grades at your university. Even if you meet the minimum requirement to apply, due to the competitive nature of the grant application process, these rewards are generally reserved for those who have a very high cumulative average, typically between a mid to high A-range. Sometimes private foundations will be more lenient with the minimum grade requirements if they do not receive as many applications, however, the major governmental institutes and councils are usually stricter with regards to their eligibility requirements. As a result, it is not worth your time or effort to apply for any research grant if your grades do not meet the minimum published requirements. If your grades are right at the minimum published requirements or better, only then is it worthwhile to submit an application. In the end, even if you do not receive the grant, you will gain excellent practice in grant writing skills.

With this in mind, you should also know that there are different strategies you can use to increase your chances of gaining a research grant, even if your grades just meet the bare minimum requirements. For example, you could focus your efforts on working within a department that does not hire as many students if the research grants administered to your university is divided up by department. To provide you with a personal example, my department of study (biology) was provided approximately twice as many research grants as the department where I was hired (medical physics) as a summer student. The number of undergraduates applying within the department of medical physics, however, was far less than half of the amount applying within the department of biology. As a result, it was more likely that I would receive the same research grant because of the department I applied within.

Another strategy is to apply for a research grant upon joining a research group within another department whose students typically score lower grades due to the nature of the subject. For example, if you belong to a department in which many students achieve 90% averages (including yourself), it may be worthwhile for you to apply for a research grant within a department in which the majority of students achieve 80% averages, as this makes you a more competitive applicant provided that grants are awarded primarily based on grades (as is often the case).

There may be other strategies that you can employ to increase your chances of being awarded with a grant, however, because these vary from university to university (and even country to country), you will need to ask the professors, graduate students or previously awarded undergraduates in your research group or field, in order to better strategize.

Beyond your grades, there are additional factors that are considered when undergraduate research grants are awarded. Because most of these factors are within your control, it is important that you put in the additional effort to ensure that you are increasing the quality of your application. Again, these factors vary from grant to grant, but it is not uncommon for you to be asked to complete a series of questions, submit your curriculum vitae, or even provide a letter of reference. If you want additional information, please refer back to Chapters 5 and 6 where I discussed how to correctly and successfully prepare these items.

The most important non-grade component of your grant application, however, is a document known as a research proposal. Unfortunately, research proposal writing is a poorly taught (if not untaught) skill in a science student's undergraduate career. While your undergraduate degree will teach you from textbooks that contain facts all compiled from and validated by hundreds, if not thousands, of research experiments, it is highly unlikely that you will learn how researchers gain the financial ability to carry out such projects. As a result, I will share all the necessary details that you will need to successfully write a research proposal for any undergraduate research grant application in the next section of this chapter.

Academic Writing in Research

Mastering the Abstract

Before you begin the process of writing a research proposal, I highly recommend that you first practise writing high-quality research abstracts. Once you have successfully mastered the skill of abstract writing, you will find that writing a research grant proposal becomes significantly easier. Besides the similarity between the two types of writing, abstract submissions are also often used in a competitive environment just like research grants. While many researchers will submit their application in hopes of presenting their projects at well-known or prestigious research conferences, only the researchers that showcase the highest-quality research and the best written abstracts will be selected to present their work.

Knowledge and Discovery

To many students, writing an abstract can be a rather daunting task, as it requires you, as the researcher, to condense months of your work into a summary of approximately 150-300 words. This takes practice as it is necessary that you distinguish between what is important to include and what is irrelevant. Based on the research conference, symposium or journal submission, each and every abstract will follow a slightly different format. It is very important that you follow the specific abstract format pertinent to where you are making a submission, as abstracts that do not follow the requested format are simply discarded.

That being said, all abstracts generally follow a standard set of rules that all applicants are required to know about before making any submissions. Unfortunately, writing an abstract is not quite as simple as providing a summary of a story book, as even though this piece is so short it needs to be written to include four main components, in addition to the initial title and author information.

First begin by writing down your title. Your title should be creative and should both accurately summarize your research project and be as short and concise as possible.

Next, provide the names of all the authors involved in your project. In general, the order of the names on your abstract will follow the same order used in an academic publication, with the exception that the presenter (you) is typically named as the first author, even if they did not necessarily contribute more work than the other authors. The remaining authors' names are arranged in order of contribution from greatest to least, followed by your co-supervisor(s) (if any), and finally your supervisor.

Some abstracts also ask for a short section on keywords used for indexing purposes. If this is the case, include the maximum number of words or phrases relevant to your project as requested by the guidelines you are following. Use words or phrases specific to your project, while avoiding overly-generic words, such as "disease" or "biology", as they are not specific enough for someone who is searching for research abstracts in a certain sub-field.

Once you have completed your author and title information as well as your section on keywords, you are now ready to write the main portion of your abstract, which consists of the following four sections: introduction (or background), methods, results and conclusion(s).

Your introduction section should contain a few sentences strictly devoted to summarizing your project's background. This means providing the reader with the interesting facts that highlight the importance of your research, without necessarily divulging any specific information about your actual project. For example, you may wish to begin your introduction with a powerful opening by stating that "More than a million people are affected by [a disease] worldwide" or "Environmental conservation is particularly important as thousands of species become endangered each year due to [anthropogenic factor]". By writing an introduction, you are explaining how your particular research project is relevant to the field of study while providing the reader with a broad background to a highly specific experiment.

Your methods section is the first portion of your abstract that contains project-specific information read by your audience. This is arguably the most difficult portion of the abstract to write because you must decide what components of the experimental protocol you wish to share with your readers, in the short amount of space provided. In this section, you should provide information about the nature of your study. You should also define the major materials and/or subjects used in your experiment, and how they were selected. Provide a description of the interventions and explain how outcomes were measured. Finally, provide a short sentence explaining the statistical methods used to analyze your collected data. You may have to edit this section multiple times in order to avoid any details that are too specific as this will take up valuable abstract space.

Your results section should begin with a description of your experimental outcomes. It is important that you also make comparisons between treated and control groups, and whether a statistically significant difference between the two groups was found. Also, explain whether any trials or subjects were excluded from the study, and if so, provide the reasons for this.

Finally, in your conclusion section, you should clearly define how and why your findings are relevant by stating what you can conclude from your study. Describe the future implications of your study results and explain why the findings of the project contribute to the growth of the research field. If you have additional space, you may also include any future prospects or considerations.

Knowledge and Discovery

Besides grant writing, honing your abstract-writing skills allows you to improve your ability to communicate your research in academia with others in the field. If you have a good understanding of how to write an abstract, you now have the general template that you will also use for constructing oral presentations, posters, and academic papers in research. Though a detailed discussion regarding how to successfully construct these items is beyond the scope of this book, I have provided you with a series of guidelines pertinent to scientific writing that can be found in Section 4 of Appendix B. In addition, before you begin writing, you should always consult the guidelines compiled by the specific academic event or journal where you plan to make a submission.

Writing an Undergraduate Research Grant Proposal

Now that you have a strong understanding of how to write an abstract, explaining the process of writing an undergraduate research grant proposal becomes much easier. At the undergraduate level, a grant proposal is usually between one and three pages long. This provides students with the opportunity to provide considerably more detail about their research project when compared to an abstract.

Though there may be some slight variance between proposals, all sections of a research grant proposal at the undergraduate level are similar to an abstract, except for one. For example, it is required that you provide an introductory section that explains the purpose of your study and relevant background information in the form of a short literature review. Next, you will write a section on your experiment's methods and design, once again similar to an abstract. Some students find the methods section easier to write in a grant proposal versus an abstract because applicants are allotted more space. The two final sections (i.e. results and conclusion), however, will likely be replaced by two new sections as follows: study justification and future prospects.

The section on future prospects is highly similar to your discussion of future considerations in the conclusion of an abstract. Here you are asked to predict ahead of time, before you obtain any results, what future experiments you wish to carry out following the completion of your current study. Because you are being asked about future prospects before you even begin your study, you do not need to be as specific as if you were writing an abstract, however, it is important that you are not too vague either. If you wish to be successful, it is necessary that the grant selection committee is able to understand the relevance of your project, but also how you plan to improve the study design in the future. As a result, you need to identify what other projects can be designed to answer any questions that your current project has the potential to generate once completed.

While all the components of a grant proposal discussed thus far are arguably detailed versions of abstract sections, justifying your study is a component that creates a unique difference between these two written pieces. Writing this section of a grant proposal is arguably the most difficult, as those involved in selecting successful applicants review your justifications closely. Funding is always a limiting resource in the world of research, and therefore, you must use this section to convince the selection panel that it is more important for your study to be funded when compared to that of a competing applicant.

While your content needs to be convincing, it is important that you understand that writing this section is not an invitation to include biased opinions and personal experiences. As a result, you should never state that your project should be funded because of any reason that is unwarranted or unsupported by the current research literature. Instead, you will need to focus on the information that exists in the scientific literature that strongly supports the importance of conducting your research project.

Good reasons that justify your study may include any or all of the following: success of your proposed intervention in past studies, a consensus that there is a poor understanding of your proposed research project (or subfield) in the scientific community, a high potential for a novel discovery or novel approach in solving a problem, a high potential for positive impact on a component of the bigger picture (i.e. human populations, animal populations, the environment, etc.), a proposal that seeks to answer a question or solve a problem that is highly urgent or important in today's research world. This is not meant to be an exhaustive list of good justifications for your study, however, what I have provided will allow you to gain greater insight into how this section is written.

Other Opportunities for Advancement

Besides obtaining grants during your undergraduate research career, there are also other opportunities for advancement that are worth mentioning. Earlier in this book, I explained different ways in which you can showcase your academic work through writing abstracts, presenting research posters or talks, or even co-authoring a scientific publication, however, mastering your communication skills is only half the battle. Next, you need to find an event or academic journal that can provide you with the opportunity to showcase your work, and this can sometimes be the phase that presents students with the greatest difficulty.

Presenting Your Project

Some of the easiest ways to gain experience presenting your work is through an event organized specifically for your undergraduate research experience. This could be the poster session that showcases the work of all the summer students at a given university department, or it could be an undergraduate symposium showcasing the work of research project or thesis students.

Presenting your work at these events is a great way to familiarize yourself with the organization and expectations associated with scientific conferences or symposiums. While you have the opportunity to practise your presenting skills in a real-world format, typically with actual judges comprised of professors and graduate students, there is no expectation or added pressure for you to deliver a flawless performance or risk your professional reputation by defending your research poorly. Instead, many judges at these events will visit your poster or attend your talk, listen to you present your work, and then provide you with constructive feedback on how you may improve.

The suggestions that you receive could be anything including changing the format of your PowerPoint™ presentation or research poster, improving the quality of your oral communication skills, answering questions you were asked in a different fashion, or even taking different approaches to ameliorate your project's research design or the actual experiment. Best of all, it is almost always the case that you will not need to compete in order to present your work at these events since they are opportunities (or mandatory requirements) provided to each student enrolled in the research program.

Another option is to pursue opportunities to present your work outside of your department at a conference or symposium that requires an application and accepts presenters on an invitation-only basis. These are ideal for you to apply to once you have presented your work at multiple events held by your local department or program, and you are more comfortable with the idea of sharing your research project with others. These events typically exist at the undergraduate and the graduate/professional level.

At undergraduate level events, all the presenters are undergraduate students, who are typically presenting a small portion of a larger project belonging to a graduate student or professor in their research group. It is less common for undergraduates to present work on their own independent project, however, this is also sometimes the case. Unfortunately, while this may help you to gain distinction within your undergraduate peers, most graduate students and professors will not consider these events to be exceptionally important from a professional standpoint.

If you wish instead to contribute to a professional level conference or symposium, you will likely need to work alongside your supervisor. These events only showcase the most recent and important research projects in the field, and it is rarer that an undergraduate would have the opportunity to contribute to a research poster or talk of this level.

Contributing to events such as these can greatly improve your credibility and level of commitment as an undergraduate researcher as your work will be showcased in front of hundreds, if not thousands, of experts within your field. Ask your professor if he or she has an upcoming opportunity, and if so, find out whether you could make a contribution.

Knowledge and Discovery

Presenting at a professional event can be challenging for undergraduates, because in order to contribute they usually need to have acquired significant findings in their research data over a time span of just a few months. Usually, students engaged in longer research terms, such as an extended commitment to a research group, or a senior thesis project, are those who are fortunate enough to gain significant findings, however, nothing in research is ever a guarantee.

I was fortunate enough to make some significant findings during my time as a senior thesis student in my final undergraduate year, and as a result, I was able to contribute to multiple professional events through writing the abstract or constructing the research poster. Other ways to contribute could include constructing an oral presentation, or presenting or co-presenting your work at the event itself, though the latter may be more difficult due to the financial costs associated with your travel expenses, especially if the conference is held in another country or even continent.

Regardless of whether the event is at the undergraduate or graduate/professional level, the application process for any of these events typically requires that you first submit your abstract for consideration. This is one of the reasons why I highlighted earlier that writing a high-quality abstract is important if you want to be successful in academia, as this is the only document many selection committees will use to assess the quality and impact of your research project. If your abstract follows all the required guidelines and impresses the selection committee, then it will be approved and you will receive an invitation to present your work at the event.

Publishing Your Work

Another way to contribute to the research world as an undergraduate is to co-author an academic publication. Again, this opportunity exists at the undergraduate and graduate/professional levels, and the benefits and disadvantages of each are similar to that of presenting a research poster or talk at an academic conference. Co-authoring an academic publication is a significantly more difficult feat to accomplish than presenting your work at an academic conference.

By now you should know that the vast majority of academic publications come in the following two forms: primary research and review articles. Co-authoring either of these articles takes significant amounts of time, effort and dedication, while the former requires the added challenge of obtaining sufficient publishable results. While you may be able to submit a small part of a project as a worthy undergraduate-level journal publication, the requirements of a professional journal are far more demanding.

Every publication begins with the culmination of an analysis of research data from an experiment reported in the form of a manuscript. Once the manuscript has been written, it is then sent off to an academic journal for consideration where it will undergo a series of edits, which sometimes spans a few months. As a result, undergraduates typically do not have the time or ability to co-author a paper due to their other commitments such as their university coursework, or if they do, the paper is published a year or two after they have left their research group, and perhaps even after they have graduated.

While it is certainly not a necessity to co-author a scientific publication, if you are one of the very few undergraduates who accomplish this feat, it is my thought that the achievement is well-worth it. Besides making a meaningful contribution to the research world at large, you can promote your article as highly sought after proof of your untiring dedication and commitment to research on your professional and graduate school applications, and can thereby increase your chances for gaining acceptance to your favourite programs.

Search Actively for Advancement Opportunities

Regardless of what circumstances exist for you to advance in your undergraduate research career at your university of study, it is important for you always remain proactive on your search for available opportunities. More often than not, and just like research experiences themselves, the vast majority of opportunities are hidden, unadvertised and obtained through networking with those who have already invested their careers and education in academia.

Knowledge and Discovery

While many keen students focus on planning for future opportunities meticulously, I wish to caution you that this may lead to disappointment. Instead, I would recommend that you keep your opportunities for advancement open, and plan instead on being open to accepting opportunities that show up in unforeseen or unexpected ways. Sometimes this leads to greater advancement than focussing your efforts on just one or two opportunities that may never come to fruition during your time as an undergraduate. Regardless of the opportunities you may be given, however, I can assure you that success in research is only reached through dedication, hard work, and most importantly, the willingness to learn.

CHAPTER 12
Remembering the Bigger Picture: Concluding Remarks

The Personal Impacts of Research

It is my hope that the strategies and personal experiences provided within this book will both encourage and empower you to actively pursue research during your time as an undergraduate. Though I need not emphasize the importance of research to you once more, I encourage you to use your experiences to shape your planning process as you begin your university education. Provided that you select your research experiences carefully, I can assure that you will benefit from them in more ways than you currently realize. This can mean that you confirm your love for a field you had an interest in since you were young, that you develop a passion for research and pursue graduate school like me, or it could even mean you come to the realization that the world of research is not suited to your interests and likings at all.

Regardless of your thoughts about research, your conclusion is a benefit in itself as you would not have been able to make one had you not involved yourself in the first place. What you will learn from gaining even one undergraduate research experience are valuable skills that you will retain for a lifetime. I know not of any other activity that you can partake in during your undergraduate career that allows you to network with and work alongside leading experts in a field that interests you. Beyond this, experience in research promotes the practical application of what you learn in your undergraduate classes, helps you to develop distinctive marketable skills, and provides you with information that may ultimately shape your future goals, education path and professional career. I also know not of any other undergraduate activity that has as great a potential to make a large and wide-spread difference in each and every issue that the world faces.

Every Contribution Counts

To put things in perspective, realize that research and development plays a fundamental role in the contribution of knowledge and discovery. In today's modern world, it is easy to see that each and every profession is shaped in some way, whether directly or indirectly by the world of research. From the way doctors perform medical procedures to help those in need and suffering, to the way engineers build the structure you live or work in, to the way educators teach the next generation of students, researchers have shaped both the profession and the professional.

Every once in a while, an undergraduate researcher makes a remarkable contribution to research, such as discovering a new species, finding a more effective method of targeting cancer cells, or engineering a device so useful and economical that it can be distributed to millions of impoverished people, and there is no reason that you could not do the same.

I am not suggesting that you should expect to make a remarkable discovery, but instead, I am ensuring that you are made aware that you hold that potential. I am suggesting that you think of each research experience as a highly-respected privilege that allows you to make a difference no matter how big or small. In other words, the result of research leads to the creation of new knowledge.

Even as an undergraduate assisting a graduate student or professor with a larger project, by conducting research you are helping along the process of creating new knowledge within your specific research group's field. Though these contributions may seem minute as you tirelessly run experiments or tediously collect data, it is important for you to never forget the bigger picture and the overall goal of research.

If your findings are published, whether through an independent experiment or a component of a larger project, this new knowledge will be read and reviewed by field-specific experts across the entire world. Eventually, with sufficient further studies supporting your research group's published findings, this information will make up a textbook used to educate the next generation of students, change the way current issues in the sciences are addressed for the better, and potentially even change the way we view the world around us.

Keeping this in mind, it is my hope that like me, you will find that the entire undergraduate research process is worthwhile. Regardless of your ambitions as a student, I wish you the best in your undergraduate research career and hope that you can find as much enjoyment as I do knowing that you are making a great and positive difference.

APPENDIX A
Sample Documents

Section 1: Cover Letter

[Today's Date]
[Your Full Name]
[Your Full Address]
[City, Province/State]
[Postal/Zip Code]

[Full Name of Professor/Administrator]
[Research Group's Full Address]
[City, Province/State]
[Postal/Zip Code]

Dear Dr. [Professor's Full Name],

My name is [Your Full Name], a [Your Year of Enrollment (i.e. 1st, 2nd, etc.)] year student at [Your University] currently enrolled in [Your Degree Program]. I am actively searching for a [Season (i.e. Summer, Fall)] research position – spanning anytime from the beginning of [Specified Month] [Specified Year] to the end of [Specified Month] [Specified Year] – in order to gain [Research Type (i.e. laboratory, clinical)] research experience. I am kindly asking if you would be able to offer me such a position to work in your [Research Group] this summer, as I see you conduct research at [Name of Research Institute/Department]. If you do not have a position available for me, I would be especially grateful if you referred me to any of your colleagues that may have a position available.

[Your Primary Field of Interest (i.e. health care, animal conservation, nanotechnology, etc.)] has always fascinated me since a young age, and I have always believed that this field is of key importance. While I have taken a range of subjects, I particularly enjoy and excel in courses in [Courses Related to Your Field of Interest (i.e. biology, etc.)] and am primarily interested in conducting [Specific Research Sub-Field of Interest (i.e. microbiology, cell biology, physiology, etc.)]. Throughout the year I have enjoyed attending various colloquia featuring the latest research of various science departments on campus as well as job shadowing [Your Intended Future Profession] when time permits.

Lastly, and most importantly, I carry prior research experience. Last summer I spent a three-month research term at the [Specified University] conducting research in [Specified Field] under the supervision of professor Dr. [Specified Professor's Full Name] of the Department of [Specified Department]. I conducted experiments that aimed to investigate [Nature of Your Research Project]. I believe, from what I have stated above, that I constitute a [Qualities that Best Describe You (i.e. conscientious, determined, self-motivated, etc.)] person, and I hope to be someone that will be of great benefit to your research work.

If you are willing to provide me with this opportunity I will sincerely appreciate it. Please do not hesitate to contact me about anything, or if you wish to be provided with any further information. I would be happy to answer any further questions you may have, especially as you do not know me personally and I would be pleased to meet you in person if you wish to discuss any concerns. I sincerely look forward to hearing from you soon.

[Your Signature (Signed in pen after printing your letter)]

Thank you kindly for your time and consideration,
[Your Full Name]
[Your Degree Program]
Level [Your Year of Enrollment]
[Your University]
Phone: [Your Personal Phone Number]
[Your University-Assigned Email Address]

Additional Information about the Cover Letter

- The font type and size has been adjusted for the publication of this book. Typically, your cover letter should be printed on standard 8 ½ x 11 in paper and should not exceed a single page.
- Always submit a typed and printed cover letter, never a hand-written one. Your signature is proof that you endorse the contents of your cover letter and should always be signed in pen, never in pencil or electronically.
- This cover letter is a template and should be used to assist you in modeling your own cover letter. Not all the information provided may pertain to you.

Section 2: Hybrid Resumé-Curriculum Vitae

Curriculum Vitae of [Your Full Name]
Updated [Specified Month and Year]

Personal and Contact Information

Full Name: [Your Full Name]
Date of Birth: [Your Date of Birth (Optional)]
Place of Birth: [Your Place of Birth (Optional)]
Citizenship: [Your Citizenship, Optional]
Professional Address: [Your Professional Address (If you have one)]
Phone: [Your Phone Number (Preferably professional, otherwise personal)]
Email: [Your University-Assigned Email Address]
LinkedIn: [Your LinkedIn® Weblink (Optional)]

Professional Summary of Qualifications

A [Your Enrollment Year (i.e. 1st, 2nd, etc.)] year Bachelor of [Your Field] Sciences (B.Sc.) student at the [Your University], with a passion for [Your Primary Field of Interest (i.e. Health Care, Animal Conservation, Nanotechnology, etc.)]. Trained in conducting [Specified Broad Skill Set (if you possess prior research experience or have taken courses in research (i.e. basic laboratory science, clinical sciences, epidemiological research, etc.)]. Proficient at conducting research in a variety of settings inclusive of [Specified Examples (i.e. libraries, online databases, the lab bench, etc.)]. Able to convey complex research findings both orally and in writing. Driven to understand and learn from the needs of others with the utmost quality and compassion. Committed to exercising [Specified Qualities to Describe You (i.e. professionalism, leadership, self-motivation, integrity, etc.)] en route to every pursuit.

Knowledge and Discovery

Education

Undergraduate Studies:

[Your University Name] | [Month and Year of Initial Enrollment] –
Ongoing | [City, Province/State]

Bachelor of [Your Degree Program]
- Graduation Expected [Your Year of Expected Graduation]
- Enrolled in Level [Your Year of Enrollment]

Relevant Degree-Specific Training (subjects listed alphabetically):
- [Your Most Relevant Undergraduate Courses (i.e. biochemistry, cell biology, general chemistry, genetics statistics, etc.)]

Research Courses (Note: Add this title if you if you have completed or are enrolled in a research course.)

SAMPLE ENTRY:

[Specified Course (i.e. senior thesis, research practicum, etc.) | [Start Month and Year] – [End Month and Year]
- [Course Code] supervisor for Level IV selected as Dr. [Specified Professor's Name], [Your Professor's Main Credentials (i.e. M.D., Ph.D., etc.)], Division/Department of [Your Professor's Division/Department] (You may also wish to list an specific titles your professor has acquired, i.e. Chair, Dean, Director, etc.)
- Senior thesis co-supervisor for Level IV selected as Dr.
- [Nature of Your Project]

Secondary School:

[Your High School Name] | [Month and Year of Initial Enrollment] – [Month and Year of Graduation] | [City, Province/State]
- Awarded [Secondary School Diploma] | [Date of Graduation]
- (You may also wish to list any distinctions associated with your graduation, i.e. enriched program, honour roll, etc.)

Awards and Distinctions
(Note: Add this title if you if you are a recipient to an award, including research grants.)

SAMPLE ENTRIES:
Deans' Honour List | [Your University Name] | [Specified Academic Year]
- Named to list for achieving a sessional average of [Grade Point Average] for Levels [Specified Year] of [Your Degree Program]

[Specified Research Award] | [Granting Council, Foundation or Institute of Research Award] | [University Where You Were Awarded] | [Month and Year]
- One of [Specified Number (If known)] undergraduate students awarded research grant of [Value and Currency (i.e. $1000 CAD]
- [Terms of Research Grant (i.e. Grant funds a 16-week summer studentship)]

Research and Career-Related Experience
(Note: Add this title if you possess research-relevant job experience or after you have gained a research experience.)

SAMPLE ENTRY:

[Employing Research Institute] | Faculty of [Specified Faculty] | [Specified University] | [City and Province/State]
[*Position Title, i.e. Summer Student Researcher, etc.)*] | [First Month and Year of Employment] – [Final Month and Year of Employment OR Ongoing]
- Conducted research under the supervision of professor Dr. [Specified Professor's Name], [Your Professor's Main Credentials (i.e. M.D., Ph.D., etc.)], Division/Department of [Your Professor's Division/Department] (You may also wish to list an specific titles your professor has acquired, i.e. Chair, Dean, Director, etc.)

Knowledge and Discovery

- [Any Personal Funding Associated with Position (i.e. Funded by undergraduate research grant from the [specified institute])]
- [Nature of Research Project]

Scientific Publications
(Note: Add this title if you have published an academic paper at the undergraduate or professional level)

SAMPLE ENTRY:

- [Authors of Publication (Formatted as follows: Last Name, First and Middle Name Initials, i.e. Ng, J. Y. {Bold or italicize your own name})] [(Specified Year).] [Article's Title]. [*Journal's Title (italicized)*]. [Volume Number(Issue Number), First Page Number-Last Page Number.]

Research Presentations
(Note: Add this title if you have presented (or contributed to) a talk or research poster at an academic conferences or symposiums.)

SAMPLE ENTRY:

[Name of Conference or Symposium] | [Location or Department of Conference or Symposium] | [City, Province/State] | [Month, Date and Year of Presentation]
- [Authors of Presentation (Formatted as follows: Last Name, First and Middle Name Initials, i.e. Ng, J. Y. {Bold or italicize your own name})] [(Specified Year).] [Presentation's Title].

Academic Conferences and Symposiums Attended
(Note: If you regularly attend colloquia and have yet to gain a research experience, you can add this title to include entries that show your research interests. This title is irrelevant once you have gained your first research experience and should thereafter be removed from your curriculum vitae.)

SAMPLE ENTRY:

[Name of Conference or Symposium] | [Location or Department of Conference or Symposium] | [City, Province/State] | [Month, Date and Year of Presentation]
- [Name of Presenter(s) and their Credentials]
- [Topic and Nature of Presentation]

Community Involvement and Volunteer Service
(Note: Add this title if you possess experiences relating to research and academia, especially if it involves your particular field of interest.)

SAMPLE ENTRIES:

[Name of Conference or Symposium] | [Specified Location] | [City, Province/State]
[*Volunteer Title (i.e. Conference Volunteer, etc.)*] | [Month and Year of Service]
- [Brief Event Information (i.e. Volunteered and attended the 1st annual [Specified Research Symposium])]
- [Brief Nature of Duties (i.e. Duties included assisting with setup of symposium, answering conference attendees' questions, etc.)]

Knowledge and Discovery

[Specified University Name] | [Specified Location] | [City, Province/State]
[*Volunteer Title (i.e. University Peer Tutor, etc.)*] | [Month and Year of Service]
- [Brief Nature of Duties (i.e. Served as a peer tutor assisting first year Bachelor of Science students with subjects including chemistry, biology, and mathematics.)]

Student Organization Involvement
(Note: Again, add this title if you possess experiences relating to research and academia, especially if it involves your particular field of interest.)

SAMPLE ENTRY:

[Name of Student Organization] | [Specified University] | [City, Province/State]
[*Organization Title (i.e. Founder, Editor, Contributor, etc.)*] | [Month and Year of Service]
- [Brief Summary of Organization (i.e. The [Specified Undergraduate Journal] is an organization funded and approved as a club by the [Specified Student Union]. Its interests lie in providing a platform for undergraduate students from across Canada to publish their research in [Specified Research Field].]

Job Shadowing
(Note: Add this title if you job shadowed a professional in a field relating to research and academia, especially if it involves your particular professional field of interest.)

SAMPLE ENTRY:

Dr. [Specified Professor's Name], [Your Professor's Main Credentials (i.e. M.D., R.N. Pharm.D.., etc.)], Division/Department of [Your Professional's Division/Department] | [Specified Location] | [City, Province/State], [Duration of Job Shadowing]

Research Safety Training
(Note: You may wish to advise your potential supervisors if you possess past training experience.)

SAMPLE ENTRY:

- Completed [Specified Training Course or Seminar (i.e. Environmental & Occupational Health Seminar)

OR simply write:

Training Records Available Upon Request

Additional Skills
(Note: Add any general and specific research-related skills that you possess.)

SAMPLE STATEMENT OF GENERAL SKILLS:

- Knowledge of general laboratory procedures, safety, and the scientific method.
- Excellent oral, written, and technological communication skills.
- Possess ability to gain new skills and learn quickly. Dedicated and self-motivated.

SAMPLE STATEMENT OF SPECIFIC SKILLS:

- Experienced in the Following Techniques: Protein Quantification and Validation, DNA Electrophoresis, Cell Culture, etc.
- Experienced in the Following Skills: Patient Interviews, Survey Administration, Database Construction, etc.

Knowledge and Discovery

Future Academic Pursuits
(Note: If your future academic pursuits are related to the research experiences you are applying to, it is wise to share this information with potential supervisors.)

SAMPLE ENTRY:

- Acceptance and enrolment in a [specified graduate or professional] program with the intent of researching [specified field or research interest].

References
(Note: As I mentioned before, it is unwise to provide potential supervisors with references within your curriculum vitae. Instead of providing a reference, you may simply provide the statement "References Available Upon Request" as shown below.)

SAMPLE ENTRY:

Reference 1:
[Your Referee's Name]
[Your Referee's Academic Rank/Professional Title (i.e. Professor, Manager, Teacher, etc.)]
Department of [Your Referee's Department]
[Your Referee's Building and Office Number]
[Your Referee's Professional Institution]
[City and Province/State]
[Postal/Zip Code]
Tel: [Your Referee's Office Phone Number]
Email: [Your Referee's Institution-Assigned Email Address]

OR simply write:

References Available Upon Request

Additional Information about the Hybrid Resumé-Curriculum Vitae

- Be mindful of whether your professor or department requests a curriculum vitae or resumé. A resumé typically should not exceed two pages, and will require you to omit unnecessary entries and additional details. Sometimes a page or content limit is imposed on your curriculum vitae as well.
- Be aware that your professor has a limited amount of time to read your curriculum vitae. Even if there is no page limit, try to omit any entries that would be unnecessary to the specific research field and position you are applying for.
- You may also wish to add a header and/or footer to your curriculum vitae, containing your name, university, date and/or page numbers. This is not shown in this template due to formatting constraints.
- The font type and size has been adjusted for the publication of this book. Typically, your curriculum vitae should be printed on standard 8 ½ x 11 in paper.
- Always submit a typed and printed curriculum vitae, never a hand-written one.
- This curriculum vitae is a template and should be used to assist you in modeling your own curriculum vitae. Not all the information provided may pertain to you.

Section 3: Letter of Reference

[Today's Date]

[Your Professor's Name]
[Your Professor's Academic Rank (i.e. Assistant Professor, Full Professor, etc.)]
Department of [Your Professor's Department]
[Your Professor's Building and Office Number]
[Your Professor's University]
[City and Province/State]
[Postal/Zip Code]
Tel: [Your Professor's Office Phone Number]
Email: [Your Professor's University-Assigned Email]

Letter of Reference

It is with great pleasure that I write this letter of reference for [Your Full Name]'s acceptance into the [specified research or academic] program.

I have known [Your Name] since [Month and Year], when s/he was my student in [Specified Course] at [Specified University] in which he scored a [Grade]. Having gained an interest in my research program and wishing to join my research group this past summer, [Your Name] was the first to take the initiative in approaching me during the summer before his/her [Specified Year of Enrollment]. Upon inviting him to an interview, I was impressed with [Your Name]'s extensive research experience under the supervision of prestigious principal investigators. This is without surprise, for [Your Name] is always prepared for the endeavours s/he pursues, having read all my recent scientific publications before coming to the interview, and knowing exactly which of my projects s/he wanted to involve him/herself in.

[Your Name] also expressed interest in familiarizing him/herself with the research techniques by suggesting s/he begin as an employee during the summer. As a result, I hired [Your Name] and am continually impressed by his/her performance. [Your Name] works very industriously, putting in long-hours and weekends devoted to his/her research project. S/He is also always willing to help out with general research tasks, and is keen to assist me in editing scientific publications.

Without question, [Your Name] possesses the characteristics of a driven student, spending ample time to learn from his/her studies. He excels in both research and academic environments and continually demonstrates a willingness to learn and grow as a person. S/He has excellent writing and critical thinking skills, an ability to work well as a team member or in a leadership role, a sharp eye for detail and a logical mind, having the ability to relate his/her own life experiences to that of others, thus forming meaningful and understanding relationships with his/her peers and supervisors. Without a doubt, it is for this reason that his/her performance showcases his/her true abilities to perform as a student.

[Your Name] is not only an accomplished student, but an exemplary and well-rounded person. S/He is one of only a few students whom I can recommend as highly. Beyond his/her performance in and out of academia is his/her dedication to his/her commitments in life. S/He is always willing to pursue new endeavours with an open and engaging mind and it is for this reason that s/he has been able to find ways to experience so many unique opportunities in his/her undergraduate career.

I recommend [Your Name] without reservations. S/He is an individual of unquestionable integrity, and one who is intelligent, adaptable, and motivated by compassion, and as a result I believe that s/he will be able to excel. I believe that [Your Name] is an excellent candidate for the [specified research or academic] program. Should you be interested in discussing anything mentioned above, or if there is anything else I can do on [Your Name]'s behalf, please do not hesitate to contact me.

Yours Truly,

[Your Professor's Signature]

[Your Professor's Name], [Your Professor's Academic Credentials]

Additional Information about the Reference Letter

- Your final reference letter should always be printed on your supervisor's departmental letterhead paper and be signed by your supervisor.
- The font type and size has been adjusted for the publication of this book. Typically, your reference letter should be printed on standard 8 ½ x 11 in paper.
- This reference letter is a template and should be used to assist you in modeling your own cover letter. In general, a letter should not exceed one or two pages in length. This template provides you with many different examples, including information that may or may not pertain to you.
- Submitting a reference letter that you have written for your supervisor should only be a last resort. Ideally, you should obtain your reference letters from supervisors who are able and available to comment on your abilities independently.

Section 4: Research Application Email

Dear Dr. [Professor's Full Name],

My name is [Your Full Name], a [Your Year of Enrollment (i.e. 1st, 2nd, etc.)] year student at [Your University] currently enrolled in [Your Degree Program]. I am actively searching for a [Season (i.e. Summer, Fall)] research position – spanning anytime from the beginning of [Specified Month] [Specified Year] to the end of [Specified Month] [Specified Year] – in order to gain [Research Type (i.e. laboratory, clinical)] research experience. I am kindly asking if you would be able to offer me such a position to work in your [Research Group] this summer, as I see you conduct research at [Name of Research Institute/Department]. If you do not have a position available for me, I would be especially grateful if you referred me to any of your colleagues that may have a position available.

If you are willing to provide me with this opportunity I will sincerely appreciate it. Please do not hesitate to contact me about anything, or if you wish to be provided with any further information. I would be happy to answer any further questions you may have, especially as you do not know me personally and I would be pleased to meet you in person if you wish to discuss any concerns. I sincerely look forward to hearing from you soon.

Thank you kindly for your time and consideration,
[Your Full Name]
[Your Degree Program]
Level [Your Year of Enrollment]
[Your University]
[Your Personal Phone Number (optional)]

Additional Information about the Research Application Email

- Note that a research application email is similar to a cover letter. Your email should be shorter as you can expect that professors typically spend less time reading an email as opposed to a letter.
- If you have no intention of submitting a cover letter, then you should include information about your interests and yourself in your email as well.
- You may also consider attaching your curriculum vitae or a short research proposal to your email as well to show an increased interest in a potential supervisor's work.

Section 5: Letter of Resignation

[Today's Date]
[Your Full Name]
[Your Full Address]
[City, Province/State]
[Postal/Zip Code]

[Full Name of Professor]
[Research Group's Full Address]
[City, Province/State]
[Postal/Zip Code]

Dear Dr. [Professor's Full Name],

I hope this message finds you well. I wish to express that I have had some concerns about my [research experience, (i.e. summer studentship, senior thesis, etc.)] with regards to your availability to supervise me and keep in contact with me this year. As you have not been often available over the past [specified time period] I have found it difficult to acquire the necessary guidance I feel I need to perform optimally on my research project. That being said, I completely understand that you have many other commitments that you must also honour. I realize that you're extremely busy with your other work outside your professorship, hence I consulted the [specified administrator/program coordinator] to discuss these concerns.

At this point I would prefer to search for another research group to join and work in a direction that is more compatible with my interests and preferences to support my research education. After very careful consideration of the circumstances at hand, and having had further preliminary conversations with [specified administrator/program coordinator] I wish to resign from my position in your research group.

That being said, I do hope that we can still maintain contact going forward. I also thank you for your guidance over this summer and for being so kind as to open up your research group to me, and for your help and guidance. Without a doubt, you have been and still are an invaluable mentor to me.

Should you have any questions regarding anything that I have mentioned in this letter, please do not hesitate to ask.

I wish you success in your research program and all the best going forward.

Yours Truly,

[Your Signature (Signed in pen after printing your letter)]

Thank you kindly for your time and consideration,
[Your Full Name]
[Your Degree Program]
Level [Your Year of Enrollment]
[Your University]
Phone: [Your Personal Phone Number]
Email: [Your University-Assigned Email Address]

Additional Information about the Resignation Letter

- A resignation letter should be submitted to your professor only after careful thought and consideration of the circumstance at hand. You should first make every effort possible to resolve any difficulties you encounter within your research group.
- The font type and size has been adjusted for the publication of this book. Typically, your resignation letter should be printed on standard 8 ½ x 11 in paper and should not exceed a single page.
- Always submit a typed and printed resignation letter, never a hand-written one. Your signature is proof that you endorse the contents of your resignation letter and should always be signed in pen, never in pencil or electronically.
- This resignation letter is a template and should be used to assist you in modeling your own resignation letter. Not all the information provided may pertain to you.

APPENDIX B
Preparatory Material

Section 1: Commonly Asked Research Interview Questions

Education

1. What is your degree program and undergraduate year?

2. What undergraduate courses have you taken/enrolled in?

3. What is your grade point average? Explain why your grade point average or grade(s) in [specified subject(s)] is/are poor.

4. Why are you applying to a research group at this institution? (If you are applying to a research group outside your own university)

Research

5. Why are you specifically interested in my research program?

6. Have you read any of our research group's publications?

7. Do you have any previous research experience? If yes, describe the nature of your past research experience.

8. Are you applying for other research positions? If so, where are you applying and in what field?

9. Are you looking for a paid or volunteer research position?

10. How long of a research term are you looking to acquire? (This question will not be asked if some research terms have fixed durations, i.e. a four month summer studentship)

11. How many hours are you available to conduct research per day/week?

12. What are you hoping to gain out of this research experience?

13. What are your personal research interests?

14. Is [Specified Discussed Project Topic] of interest to you?

15. When are you available to meet the members of our research group?

Personal

16. Tell me about yourself.

17. Describe your work ethic and/or personality.

18. What are your greatest strengths/weaknesses?

19. Tell me more about [specified entry in your curriculum vitae] and what you gained from the experience?

20. Do you have any other commitments besides academic coursework?

Communication

21. Do you have good written and/or oral communication skills?

22. Do you have experience in preparing research presentations or presenting research?

23. Do you have experience in research grant writing? (If a grant opportunity exists for you)

24. Do you have experience in editing/writing scholarly articles? (If an authorship opportunity exists for you)

Requests

25. Do you have a copy of your academic transcripts?

26. Do you have a copy of your curriculum vitae?

27. Could you provide me with an academic writing sample written entirely by you?

28. Do you have a copy of your training record?

29. Could you provide me with the name of an academic reference who could comment on your abilities?

Goals

30. What are your plans for the future?

31. What are your professional/academic goals?

32. What is your ideal future career?

33. Do you plan to conduct research in the future?

34. Do you plan to attend graduate school following your undergraduate degree?

Other

35. Do you have any questions for me?

Note: This list of questions is not exhaustive, but instead meant to serve as a preparatory tool for your research position interviews. Be aware that you will unknowingly provide the answer to many of these questions without them being asked, and that the vast majority of these questions will not be asked during your interview.

Section 2: Questions for Potential Supervisors

1. **What are my responsibilities as your research student?** Ask about your exact tasks and time commitment, and ensure that you understand them both. This means finding out what aspects of the research project you will be working on, and on what day and time you will be competing your tasks. This also means asking about when research group or laboratory meetings are held, and how often you will be able to discuss any questions or concerns you may have with regards to your research project with your professor.

2. **Who will be primarily supervising my project?** Find out who you will be working with. Many professors will allow a graduate student to supervise and mentor an undergraduate, so if this is the case, ask your professor if you can be introduced to his or her research personnel.

3. **What are your expectations of me?** Make yourself aware of what your professor will expect of you in the future if you accept the position. For example, you should know whether your professor expects you to write a report, present a research poster or provide an oral talk about your project towards the end of your research term. Perhaps if you are applying to a summer studentship, the department will require that you must present your work at a symposium in a certain format, and so you should ensure that you are aware of your commitments with regards to presenting your research.

4. **Are there any opportunities for advancement in your research group?** Identify what opportunities exist for your benefit. You may also wish to ask whether there are any research grant or scholarship opportunities available for you to apply for or whether you are able to gain authorship on any of the projects that resulted in an academic publication.

5. **What orientation sessions do I need to attend?** Inquire about the safety training and site-specific training that you will be required to complete. This is often a component of your research position that your professor will forget about, however, university regulation usually demands that all research personnel complete training courses in workplace safety. If you are working with anything that may present you with additional hazards, such as laboratory chemicals, animals or radioactivity, you may be required to enroll in additional hazard-specific courses. You may also need to acquire some site-specific skills, regarding the operation of certain equipment or the learning of certain experimental protocols. Regardless of the nature of your research position, it is very important that you possess all the required training certifications or you could face disciplinary action by the academic institution even if you were not aware.

6. **Is this a paid or voluntary position? If this is a paid position, what is my financial compensation? If this is a voluntary position, is there any opportunity for a paid position in the future?** Confirm all things financial associated with your position. One important clarification that many students are afraid to ask about is their salary. Make sure you are very clear as to whether the position you are about to accept is voluntary or paid by confirming this with your professor. If you are being offered a volunteer position, you may want to politely ask whether your professor would consider a paid position in the future. As long as you clarify this in a polite manner, there is no reason why this should offend your professor. It is certainly not a good idea for you to assume what financial compensation you may receive, and it is even worse if you have to discuss this with your professor half-way through your research experience. Remember that it is better that you clarify misunderstandings during your interview, than find something to be unexpected after accepting the position.

Section 3: Guidelines for Presenting Your Research Poster or Talk

1. Begin early. Preparing any research poster or talk at the last minute greatly increases your chances of including errors. Aim to present your work in the format of a story, and starting with an explanation of how and why you began your project and ending with the results, conclusions and future implications derived from your experiment.

2. Know your audience. If you are presenting to a symposium inviting many undergraduates, you may need to provide a more in-depth background to your research. If your audience is comprised of professors and graduate students in the same research field, you should focus on presenting less background information and more experiment-specific details.

3. Always practise presenting your research poster or talk before attending the event. Ask friends, colleagues and your professor to hear you present, if this is an available option, and ask them to ask questions and provide feedback.

4. If you are presenting a poster, ensure that the size of your text is large enough for your audience. When you view your electronic file at 100%, you should be able to read the text clearly when standing three feet back from your computer.

5. If you are presenting a PowerPoint™ presentation, be aware of the projected screen size, and the distance between the screen and the last row of your audience.

6. Be mindful of your colour contrast. Colours that do not mesh well together can easily ruin a PowerPoint™ presentation or a research poster. In general, use light-coloured text on a dark-coloured background or vice-versa.

7. Keep all your figures and diagrams as simple and clean as possible, so that they are easily understandable and visible to your audience. This applies to both a PowerPoint™ presentation and research poster.

8. Be mindful that print shops may need up to a week in advance to print your poster. Factor this in when planning how much time you have to construct your poster.

9. Be mindful that some conferences run on tight schedules, and thus, require you to submit your final PowerPoint™ file to them in advance of the actual presentation date, with no further option to edit your file. Factor this in when planning how much time you have to construct your presentation.

10. Try to include as few words as possible when constructing your talk or poster. Your audience will focus on the quality of your presentation, and will less often read the text components of your talk or poster. Instead, include images and charts to showcase your experimental protocols and results.

11. If you are giving an oral talk, avoid the use of unnecessary animations and sound effects in your PowerPoint™ presentation. This is often very distracting to your audience and takes away from your main message.

12. If you are giving an oral talk, do not speed through slides. A professor of mine once advised me to cover one slide per minute. As a result, if you are providing a 10 minute talk (excluding time for questions), then you should have 10 slides of content.

13. Never exceed the time limits imposed on your oral presentation or poster by conference guidelines, as this shows poor planning. Never present a talk that is too short either, as this indicates that you do not have enough important work to share. Be mindful of this and time your presentation when practising.

14. If the duration of your presentation is 15 minutes, you should aim to speak for 10 or 11 minutes and save the last four or five minutes for questions. If by misfortune, you realize that you will be unable to cover all of your PowerPoint™ slides within the allotted time period, skip the least importance ones, instead of beginning to talk faster. Presenting information containing less quantity and more quality is always better than the vice-versa.

15. A good presentation, regardless of whether it is a poster or talk, should induce questions from your audience. Allow time for a question period to clarify any concerns, address questions or share any additional information.

16. Do not be intimidated by the fact that you may not know all of the answers to the questions you are asked. It is acceptable to admit that you do not know an answer to a difficult question and that you will get back to the person at a later time. It is not acceptable to "make up" an answer to a question you do not know.

17. Be aware that your audience members are voluntarily spending their time listening to your talk or presentation. Be sure to thank them for their time and interest in your work.

Section 4: The Foundations of Scientific Writing

The Audience

1. When writing anything, it is important that you are aware of your audience. Different audiences will have very different levels of knowledge about your subject area (i.e. general audience vs. professionals in the field), so provide as much or as little background information based on your type of audience.

The Title

2. Your title should be written as clearly and concisely as possible, so as to reflect the purpose and/or results of your study.

3. Typically, your title in any piece of writing needs to be a full sentence that highlights the main result obtained. Alternatively, your title could be a detailed question that specifically identifies the aim of your study.

The Abstract

4. An abstract can be thought of as a summary of your research project. It is important that your abstract is well-written as this small paragraph is what your scientific audience will use to determine whether they wish to continue reading or not.

5. Your abstract should meet the word restriction specifications set by the academic event or journal at which you are making a submission. If there are no specifications, you should aim to write an abstract of approximately 300 words. If it is too long, it will contain too many unnecessary details, and if it is too short it will not sufficiently summarize your study.

6. An abstract should aim to answer the following four questions:
   a. "What is the purpose of the study?" (Introduction/Background)
   b. "How did you conduct the study?" (Methods)
   c. "What were your findings?" (Results)
   d. "How can these findings be interpreted?" (Conclusion)

7. An abstract is usually presented directly following your title. For this reason, avoid repeating the title within your abstract.

The Style and Formatting

8. Always provide the full word or phrase before introducing an abbreviation. Show that you will be abbreviating a word or phrase for the remainder of your written work by placing the abbreviation in parentheses immediately after the first use of the full word or phrase.

9. Avoid including language that is deliberately obscure or unnecessarily complex. Try your best to use language that is simple, clear, concise, yet professional.

10. Avoid the use of informality, which is common in how you communicate in everyday life. This means avoiding the use of colloquial language (slang), contractions, excessive use of exclamation, or inappropriate words or phrases.

11. In general, you should write in a passive voice as opposed to an active voice. Example: Researchers combined Substance X with Substance Y (Active). Substance X was combined with Substance Y (Passive).

12. Avoid the use of first-person pronouns in any piece of academic writing (i.e. I, we, etc.).

13. In scientific writing, the two most common tenses are present and past. You should write in present tense when making an assertion or explaining a concept, while you should write in past tense when you are explaining what has been asserted or

what protocols were followed during your experiment. Future tense is rarely used, and is used to explain future plans or implications.

The Components

14. A typical experimental report should contain the following components: title page, abstract, introduction, methods and materials, results, discussion, conclusion, references, appendices. As the requirements for each of these sections for each academic event/journal are slightly different, it is imperative that you read the specific guidelines pertinent to where you are making a submission very carefully.

15. Writing the methods section can often present difficulty. Remember to provide information about the nature of your study such as whether your project is a laboratory-based study, a clinical trial, etc. You should also define the major materials and/or subjects used in your experiment, whether they are bacteria, stem cells, mice, humans, etc., and describe how they were selected (i.e. only male mice with a genetic predisposition to diabetes were selected, etc.). Finally, provide a description of the interventions (i.e. the cancer cells were treated with [specified amount] of [chemotherapeutic compound] each day over a one week period), and explain how outcomes were measured.

The References

16. Your references should contain all the works you consulted to complete your written piece. This can range anywhere from 5 references on a research poster or talk to over 100 references on an academic publication.

17. Be sure to use the correct referencing style based on the subject of your writing. Typically, research in the natural sciences follows the American Psychological Association, while research in the clinical sciences follows the Numbering System, though this can vary from journal to journal or based on your supervisor's request.

The Appendices

18. Your appendices should contain calculations, figures, tables, graphs and anything else that is not contained within the main written piece. Be sure to refer to these items within your written piece, and be sure to correctly label/title them in your appendices.

INDEX

D

E

F

ABOUT THE AUTHOR

Jeremy Y. Ng is a current graduate student at the Leslie Dan Faculty of Pharmacy at the University of Toronto, and earned his Bachelor of Science in Biology, and graduated with distinction *summa cum laude* at McMaster University in Hamilton, Ontario, Canada. Throughout the time he spent as an undergraduate student, he was able to acquire multiple research experiences that ultimately helped shape his decision to pursue graduate studies. Jeremy has been involved in research since his freshman year, and has since worked alongside seven different principal investigators, including distinguished research chairs and clinician-scientists. Jeremy's primary research interests are in healthcare and medicine, with a specific interest in assessing the efficacy of theories and therapies associated with the complex and poorly-understood field of complementary and alternative medicine. He is the recipient of the prestigious Undergraduate Student Research Award funded by the Natural Sciences and Engineering Research Council of Canada (NSERC), as well as one of two students nationally to be awarded the KNOR Knowledge Research Grant funded by the KNOR Foundation for Cancer Research in 2012. Jeremy has shared his experiences in research through his roles as a teaching assistant, student organization leader, and undergraduate and graduate student, encouraging students to begin their undergraduate research career as early as possible in order to maximize their future potential in academia.

NOTES

NOTES

NOTES

NOTES

www.ingramcontent.com/pod-product-compliance
Lightning Source LLC
Chambersburg PA
CBHW072218270326
41930CB00010B/1906